*O*il and gas equities: evaluation and trading

*O*il and gas equities: evaluation and trading

Nick Antill and Robert Arnott

International Equities Series

WOODHEAD PUBLISHING LIMITED

Cambridge England

Published by Woodhead Publishing Limited, Abington Hall,
Abington, Cambridge CB1 6AH, England

First published 1994, Woodhead Publishing Limited

British Library Cataloguing in Publication Data
A catalogue record for this book is available from the British Library

ISBN 1 85573 119 3

Designed by Andrew Jones (text) and Chris Feely (jacket).
Typeset by Best-set Typesetter Ltd, Hong Kong.
Printed by St Edmundsbury Press, Bury St Edmunds, England.

For J.A. and J.T.

Contents

*P*reface

Equity markets value shares in companies by establishing the clearing price at which buying and selling interest is matched. They do this as part of their role as secondary markets, that is, markets in which existing shares change hands between investors. In addition to providing liquidity, however, this role relates directly to the primary function of an equity market: it sets the price at which a company can raise new equity capital, or at which it can be acquired.

Market value is thus set by investor behaviour. Economic theory suggests that in a liquid market (one with a large number of active and well informed buyers and sellers) the price which the market establishes will accurately reflect the balance of risks and returns which the share offers investors, in relation to those offered by alternative assets. If the equity market is efficient in this sense, it should be possible to relate the way in which the market prices assets to economic evaluation of the returns that they offer. All investment analysis is thus an exercise in applied economics. The models used represent an attempt to explain and to predict the prices at which shares find a market equilibrium. This book is an attempt to present the information and the theoretical knowledge required to analyse shares in oil and gas companies.

Investment analysis is entirely concerned with the future, in that it is the future returns which will accrue to a shareholder as a result of his (or her) owning a share which will determine the price which he is prepared to pay for it now. Expectations for the future are formed by interpretations of past experience, however, and in practice a range of factors will influence the process of share price appraisal. These include assessments of a company's historical performance and of its apparent intentions for the future, appraisal of the prospects for its areas of operation, as well as simple evaluation of its assets as if they were about to be broken up and sold. Companies can and do add to or subtract from their value by management actions.

This book begins with a brief description of the oil industry, its history and its ownership. It examines the external factors, notably of course the crude oil price, which combine to create the environment in which oil companies operate, and discusses the considerations which should underlie assumptions on which oil companies are evaluated. It discusses oil industry accounts, to highlight the main points of which the analyst needs to be aware when interpreting them. Finally, it discusses methodologies for evaluation of both exploration and production companies and integrated oil companies.

The book is intended mainly for those involved with the equity market, whether as investors or in the treasury or planning department of oil companies. It may also be of interest to students of business studies or energy economics. In essence, it sets out to establish a series of benchmarks and performance indicators against which the market should (and does) evaluate shares in oil companies.

There is thus no contradiction between the use of objective methods of valuation, on the one hand, and the recognition that value is a function of investor behaviour, on the other. It is necessary to remember, however, that the purpose of the models is to predict the behaviour of the market, not the other way around. As with all economic theory, techniques for share price evaluation are only as good as their predictive accuracy.

Acknowledgements

During the course of researching and writing this book the authors have benefited from conversations with a wide variety of individuals, including employees in the oil industry and in the investment community. Specific acknowledgement is due to help and advice received from Neil Atkinson, Peter Norris, Judith Parry and Brian Williams, and for permission to reproduce material supplied by BP, Petroleos de Venezuela and the Royal Dutch/Shell group. The authors remain, however, fully responsible for any errors which the book still contains.

*L*ist of abbreviations

bcf	billion (thousand million) cubic feet
bcm	billion (thousand million) cubic metres
bc/d	barrels of condensate per day
boe	barrels of oil equivalent
boe/d	barrels of oil equivalent per day
bbl	barrels of oil
bbl/d	barrels of oil per day
btu	British thermal unit
LNG	liquefied natural gas
LPG	liquefied petroleum gas
mbbl	thousand barrels of oil
mbbl/d	thousand barrels of oil per day
mboe/d	thousand barrels of oil equivalent per day
mscf	thousand standard cubic feet
mmscf	million standard cubic feet
mmscf/d	million standard cubic feet per day
mmbbl	million barrels of oil
mmboe	million barrels of oil equivalent
Sm3	standard cubic metre

CHAPTER ONE

The development of the oil industry

INTRODUCTION

It is not the intention of this book to cover in any great depth the development of the oil industry since many good texts on the subject exist already. However, for the analyst coming into the sector for the first time it is useful to have at hand a very brief résumé of the history of the industry. This chapter is therefore divided into four main sections: a brief history of the oil market, the origins of the major companies, a review of the worldwide occurrence and ownership of hydrocarbons, and a summary of the state of the oil and gas industry today.

OIL INDUSTRY HISTORY

The history of the oil industry is not only linked to each major technological invention but is also intimately linked with major political events. Briefly, the history can be subdivided into three main periods:

- The early history, lasting until Drake's oil discovery well in Pennsylvania in 1859.
- The period from 1859 to 1955, during which time more than half of the world's crude oil production came from the USA.
- The recent history from 1955 and the emergence of the OPEC period, when oil prices were dominated by a single cartel.

Early history

Oil products have been a useful commodity since around 5000 BC. The Middle East civilisations used it as a covering for boats and dikes and the Chinese are reported to have used it to illuminate the palace of their emperor. Its use as a weapon was recorded in the East Roman Empire in 650 AD, while in medieval Europe oil was used for medical purposes, lubricants and paint fabrication. Competition is mentioned in 1430 between oil producers in Modena (Italy) and Tegernese (Germany) and later, in 1643, Indonesian oil entered the European market. By 1800 lamps for illumination were already widely used, thus creating a high demand for lamp oil. This lamp oil was, until 1859, mainly derived from relatively expensive animal and vegetable oil which cost $32 per barrel (bbl). However, after that time the main source of lamp oil was kerosene, derived from oil seeps and somewhat less expensive. The high demand for kerosene at that time prompted the search for a cheap supply of oil, which was finally delivered after 'Colonel' Edwin Drake successfully drilled for oil in Titusville, Pennsylvania in 1859. With this well commenced a long period during which the US was to dominate the world oil industry.

1859 to 1955

Drake's first oil was sold at $19.75/bbl in January 1860. His success attracted so many to follow his example that an oversupply situation resulted, and the price dropped dramatically to $0.10/bbl in 1862, only two years after Drake's initial success. This low price in turn stimulated demand for the product and by December 1864 the price had risen sharply back up to $11/bbl. These large price fluctuations were typical of the early period of drilling in the US.

In 1878 the invention of the oil stove was to have an important effect on the petroleum industry. The stove became a commercial success following the World Fair in Paris, leading to a sharp increase in the demand for fuel oil (see Fig. 1.1). However, in the US the increased demand for oil products was more than offset by the increased supply and by 1895 the US was able to export up to 44% of its crude oil production.

Around the turn of the century this situation was to change. From 1900 to 1910 the consumption increased at 16% per annum, whereas the production increased at 12% per annum. The rapid public acceptance of the automobile after the introduction of the Model T Ford in 1908 and the First World War was the cause of the US becoming a net oil importer around 1920. This shortage proved only temporary, partly

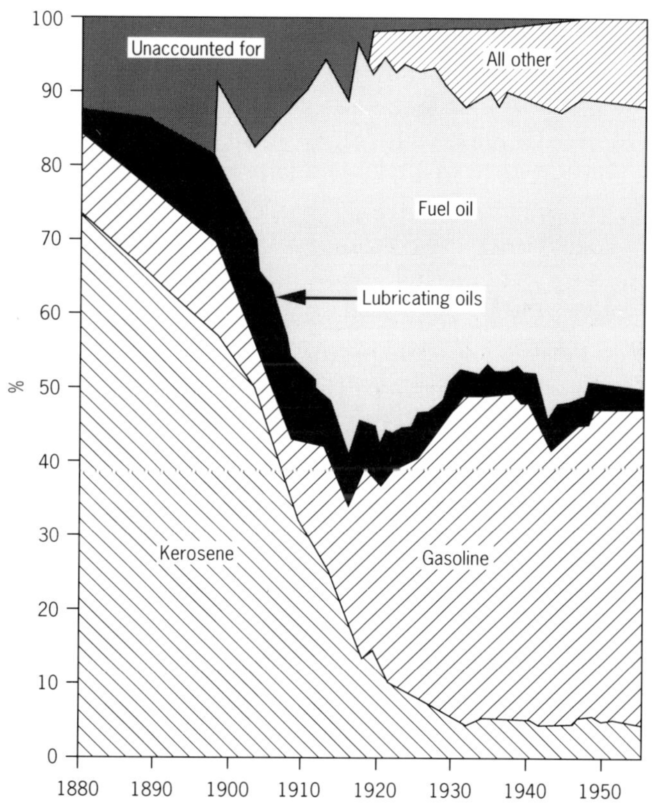

1.1 *Refinery products as percentage of crude input (source: Schurr and Netschert, 1960).*

because of the end of the war in 1918, but also because of a series of major discoveries in the early 1920s (Burbank, Smackover, and Big Lake), culminating in the discovery of the giant East Texas Field in 1930. Just as the shortage caused the oil price to increase from $0.55/bbl in 1914 to $3.00/bbl in 1920, the subsequent oversupply situation caused a price decrease to $0.65/bbl in 1931.

To avoid a complete collapse in the price of oil the US government introduced a prorationing of production in 1931, resulting in a stable price of just over $1.00/bbl from 1934 to 1945. Due to this regulation major events such as the nationalisation of the Mexican oil industry in 1938 and the Second World War were not reflected in the crude oil price.

Although the US had been importing crude oil from 1915 to 1932, the country was a net oil exporter because the export of refined products exceeded the crude oil imports. After the Second World War US consumption outpaced production and in 1947 the country once again became a net oil importer.

The US period of domination ended in 1955, from which time onwards it produced less than 50% of the world's entire oil production (see Fig. 1.2); the US remained the world's largest producer until 1964, when the Middle East took over the role. By 1984 US oil production made up only 18% of total world production.

Recent history

The decline of US oil dominance was foreshadowed by the enormous discovery rate elsewhere in the world, particularly in the Middle East, which had commenced in the mid-1930s and was to last until the mid-1960s. In 1950 the Middle East produced 1.8 million barrels per day (mmbbl/d) or 17% of the world's production. Apart from an interruption in 1951 due to the nationalisation of Iranian oil interests, this production increased steadily, and had reached 5.2 mmbbl/d (24% of world production) by 1960. To protect its oil industry from the resulting oil price decrease, the US introduced mandatory import quotas, effectively limiting the import of Middle East crudes. This move kept the oil price in the US more or less constant during the following decade, but resulted in the accelerated decline in the price of non-US crudes, since a major part of their market had become inaccessible. In September 1960 the five major oil exporting countries, Venezuela, Kuwait, Saudi

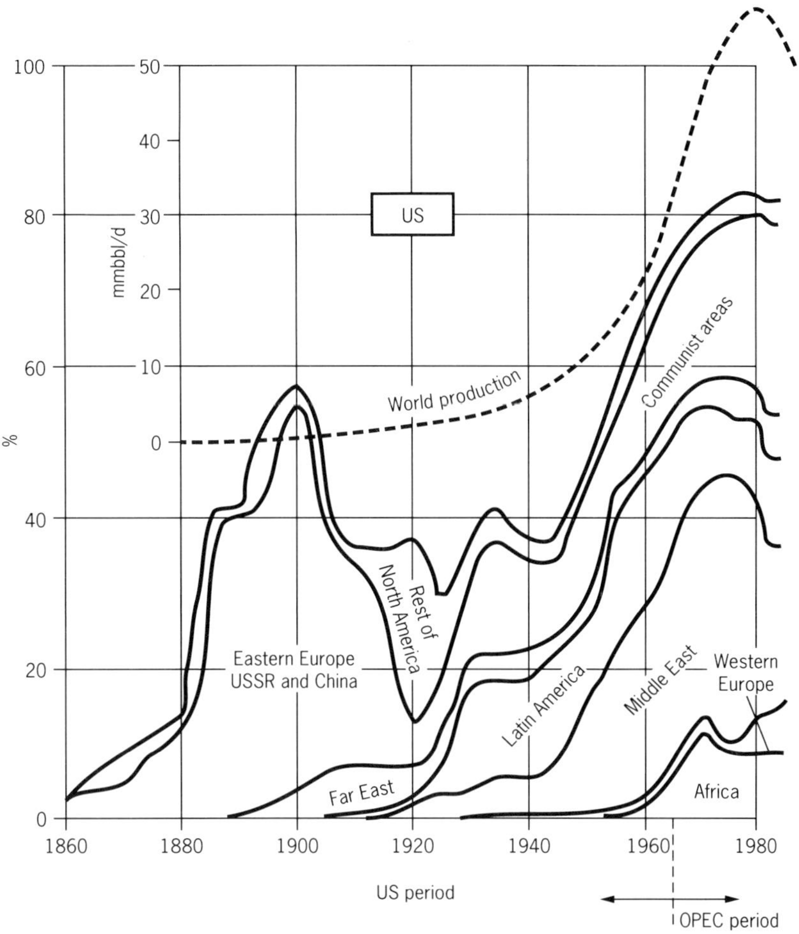

1.2 *120 years relative production rates.*

Arabia, Iran and Iraq, which together traded 85% of the world's exports at that time, created the Organization of Petroleum Exporting Countries (OPEC) in order to try and stem the price cuts and restore the price of oil.

Because of the tremendous oversupply situation the cartel had little effect initially and the oil price kept decreasing. However, the low oil prices (below $20/bbl) also induced the West European nations to switch from coal to oil. In 1960 oil supplied 29% of Europe's energy requirements but by 1970 this had increased to 56%. A similar development also took place in Japan and together both areas became heavily dependent on imported cheap Middle East crude.

In the same period, the US had seen its domestic production peak in 1968, and the country had to import 5.7 mmbbl/d – or 40% of its consumption – in 1970. The prorationing of production had been discontinued and in the spring of 1970 the US was forced to relax its import quotas.

This event triggered a set of OPEC price and tax increases which resulted in the Teheran Agreement of February 1971. The Saudi Arabian Light was set at $2.2/bbl and the tax rate went up to 55%. One year later the oil price was made independent of the strength of the US dollar in the Geneva Agreement. However, the main event was still to come. It was triggered by the Yom Kippur war of October 1973 at which time the oil price was doubled with immediate effect. The subsequent boycott by the Arab nations of the US and the Netherlands caused a panic in the spot market, which in turn triggered a new official price increase to $11.6/bbl from the beginning of 1974 (see Fig. 1.3).

The developments of 1974 were followed by a relatively quiet period when oil prices increased modestly in money of the day (MOD) terms but actually decreased when corrected for inflation. The OPEC production rate showed a temporary decrease in 1974 but recovered to its all-time high in 1977 at 31.3 mmbbl/d.

With oil prices as high as they were, it became attractive to explore and develop reserves in areas which had previously been uneconomic. As a result, there was major investment in new provinces in Alaska, Mexico and the North Sea, each of which started to produce in the late 1970s. The consequent dramatic decrease in OPEC production after 1977 was therefore a direct result of these developments, and of a continuing energy conservation campaign in the developed countries.

The combined cash flow of the OPEC members had been substantial in the period from 1973 to 1979 (some $161 billion in total), but their expenditure was increasing dramatically as well and by 1978 OPEC experienced a cash deficit of some $2 billion.

The temporary oil shortage due to the Iranian Revolution in late 1978 drove the spot price above $25/bbl in February 1979, and this triggered another round of oil price increases by the cash-hungry OPEC members. Saudi Arabia tried to moderate this by producing at high rates to suppress prices, but it could not prevent African OPEC members pushing their prices in excess of $40/bbl.

The 1979 round of OPEC price increases turned out to be poorly timed in a period when world demand was falling and many new oil fields outside OPEC were coming on stream. This oversupply situation was exacerbated by the mild winters of 1982 and 1983 in Western Europe. The oil importing nations were not obliged to buy the expensive OPEC crudes anymore and OPEC had to accept the role of 'swing producer'.

By early 1985 OPEC production was around 15 mmbbl/d – less than half of its peak production in 1977 – and its prices were under severe pressure. This was the result of the steady increase in:

- Non-OPEC supply from 20 mmbbl/d in 1979 to 25 mmbbl/d in 1984 (North Sea, Mexico, etc).
- A considerable increase in exports from the Eastern Bloc.
- A fall in world demand in the same period of some 6 mmbbl/d.

OPEC members found themselves in the squeeze and were no longer prepared to cut back production even further. During 1985, one OPEC

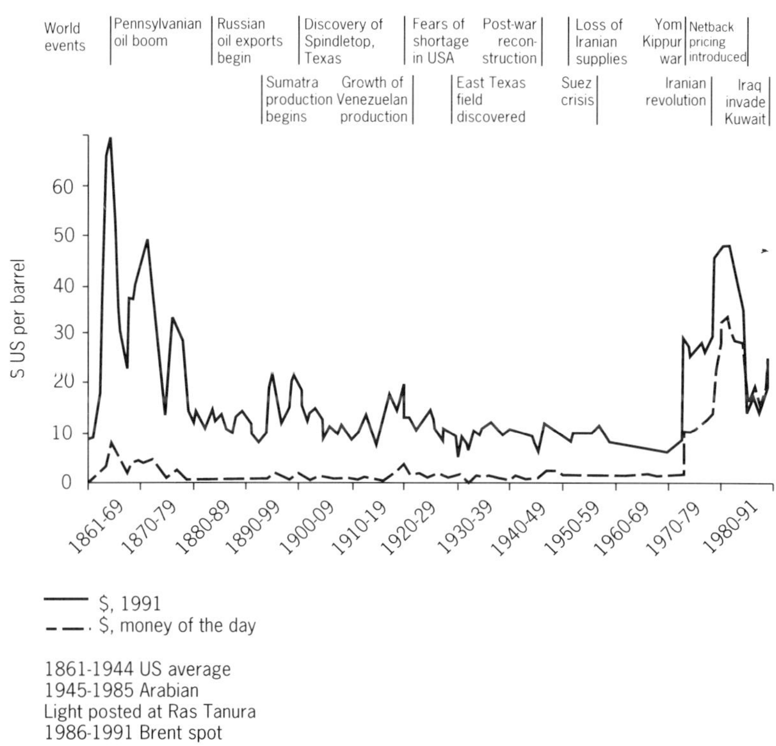

1.3 *Crude oil prices since 1861 (source:* Statistical Review of World Energy, *British Petroleum, 1992).*

member after another started to produce in excess of their agreed quotas and by the end of 1985 Saudi Arabia was no longer prepared to be the only member of OPEC to cut back on its production in order to maintain the crude oil price. As a consequence, Saudi Arabia started to increase its market share by selling crude at a netback price, thereby guaranteeing refiners a decent margin.

By the beginning of 1986 OPEC production had increased to 18 mmbbl/d, with a dramatic effect on the price of crude oil prices, which collapsed to below $10/bbl. The effect of the falling oil prices had a serious effect on the current accounts of the heavily populated oil producers whose revenues fell sharply. The main beneficiaries were oil importing countries such as Brazil.

Since 1986 OPEC has attempted to maintain the oil price flat in real terms at the level of the full cost of non-OPEC supply. To date, this has been quite successful but much will depend in the future on the long run marginal cost of non-OPEC production, especially if large additions to capacity are made (see Chapter 2). The importance of OPEC in the world oil market today cannot be overstated and the more recent history of the cartel and its impact on crude prices is discussed at length in the next chapter.

THE COMPANIES

The origins of the major oil companies run very much in parallel with the technological and political events outlined earlier. In this section, the historical development of the oil industry is reviewed briefly, with a look at the formation of the oil majors, the state companies and the more recent emergence of the oil independents.

The majors

Standard Oil and its successors

Most of the oil majors had their origins in the US in the aftermath of the Drake discovery well in 1859. The first major player was Standard Oil, formed in 1870 with the backing of John D. Rockefeller. Realising the importance of oil as a fuel, Standard Oil embarked upon an aggressive campaign of acquiring local refineries and tied up as much of the US oil industry as was possible, often cutting prices to undercut and drive competitors out of business. Such was the rapid growth in the company that over 90% of kerosene was passing through Standard Oil's facilities by the end of the 1870s.

As a result of the success of Standard Oil, there was great concern in the US over the effective monopoly that it was exerting over the oil industry. An attempt was made to break the company up but this only resulted in the formation of the Standard Oil Trust, which in effect had under its control a number of separate companies which were all still within the same group, e.g. Standard Oil New Jersey, Standard Oil Ohio, etc. The company carried on under this corporate structure until 1911 when the US Supreme Court ordered the dissolution of the Standard Oil Trust. The break-up of Standard Oil left behind seven new companies – some of which are still major players today, namely:

- Standard Oil of New Jersey which became Exxon;
- Standard Oil of New York which became Mobil;
- Standard Oil of California which became Chevron;
- Standard Oil of Ohio which became Sohio (before being taken over by BP);
- Standard Oil of Indiana which became Amoco;
- Continental Oil which became Conoco;
- Atlantic Oil which became Sun Oil.

Of all these companies the largest was Standard Oil of New Jersey which was worth over 50% of the total net value of Standard Oil. The next largest was Standard Oil of New York, worth around 9% of the total. With its head start Standard Oil of New Jersey never lost its lead and, as Exxon, it remains (alongside Shell) the largest integrated oil company in the world today.

The Royal Dutch/Shell Group

Despite the early dominance of the US oil industry another major player to emerge at the beginning of the twentieth century was Royal Dutch/Shell. This company had its origins in two separate companies:

- Royal Dutch;
- Shell Transport and Trading.

Shell Transport and Trading was set up in the late nineteenth century by Marcus Samuel, the son of a 'shell merchant'. Marcus Samuel's great strength was in spotting business opportunities and his most successful venture was to transport kerosene, in bulk, to the relatively under-exploited but also rapidly growing Far Eastern market. Such was the success of this venture that the business was exceedingly profitable whilst there was a sufficient and secure supply of oil to transport. In the early years of business most of this oil was supplied from the Russian oil fields. However, as the security of supply from the Russian fields diminished Marcus Samuel urgently required another source of supply. After several attempts he succeeded in penetrating the US market and secured a source of supply from several southern US discoveries, e.g. Spindletop. However, even these supplies had almost been exhausted by 1905 and it was around this time that serious negotiations were initiated by Royal Dutch with a view to forming a joint marketing company.

Royal Dutch had its origins in the Dutch East Indies where for several years oil seepages had been reported. By 1885 the first successful oil well had been drilled and the Royal Dutch company was launched in 1890. By 1892 the company was producing oil and within two years production had increased sixfold. However, by the turn of the century it was clear that there was need for the amalgamation of Royal Dutch and Shell if both businesses were to survive take-over attempts by Standard Oil. The union was finally cemented in 1907 when both Royal Dutch and Shell became holding companies, with Royal Dutch holding 60% of the stock in the operating subsidiaries and Shell 40%. At the time of the amalgamation in 1907 the world oil market was dominated by the original oil giant Standard Oil and the Royal Dutch/Shell Group. Even today, Royal Dutch/Shell and Standard Oil (in the form of Exxon) remain the two largest private oil companies in the world.

Other major companies

The rapid growth in US oil exploration activities at the end of the last century resulted in the growth of a number of smaller private companies, many of which still exist today. Successful drilling in California in the late 1880s resulted in the formation of Union Oil. In addition, further success in the southern US states gave rise to companies such as Gulf Oil, Sun Oil and Texaco. All of these companies were able to compete alongside Standard Oil.

Oil exploration was also active elsewhere in the world and it was at the beginning of this century that the first geological surveys of Middle Eastern countries were taking place around oil seeps. By 1909 there was real conviction of the oil potential in Iran and as a result the 'Anglo Persian' company was set up. Shares in this new venture were very much oversubscribed but by the beginning of the First World War there had been very little success. Such was the state of the company's finances by this time that it desperately needed a further injection of cash. This was, fortunately, supplied by the British government who were keen to secure a supply of oil with which to fuel the Navy (which had previously been fuelled by coal). The British government took up 51% of the company's enlarged share base and this new company was eventually (in 1954) to become British Petroleum (BP).

'The seven sisters'

The rapid growth in size of nearly all of the companies described up to this point owed much to the tremendous discovery rates that were being achieved worldwide throughout the first half of this century combined with the tremendous increase in demand which had resulted from new technology. Such was the growth in a number of the companies that the term 'seven sisters' was coined in the 1950s to describe the cartel of the major companies, which comprised Exxon, Mobil, Chevron, Texaco, Gulf, Royal Dutch/Shell and British Petroleum. The term was started because of the close association and multiple joint ventures that they were all involved in around the world. Together these companies dominated the world oil industry but their power today has somewhat diminished because of the trend in the 1950s towards nationalisation.

The nationalised industries

Nationalisation

The original players in the oil industry were all backed by individuals and institutions who were seeking to maximise profit and win concessions – and so reserves – in countries in order to secure supplies of oil. By the 1930s it was clear that the cost of producing oil in countries outside the US where major new discoveries were being made was significantly lower. Therefore the profit per barrel on these discoveries was much higher for the same oil price. At the same time, there was a worldwide trend of rising nationalism in many of the countries in which many of the discoveries had been made as the era of colonialism was drawing to an end. With a growing resentment of the large profits that were being made by foreign companies with national resources it was inevitable that the industry would not survive intact for long.

The first shots were fired by Iran (Persia) in the early 1930s when the Anglo-Persian agreement was cancelled. However, a new agreement was ratified the following year. The first real incident came in 1938 when the Mexican government nationalised the oil industry and did not compensate any of the foreign companies (most of them American) that had participated. Throughout the 1940s and 1950s host governments began to tighten the grip on their own resources – including most of the Middle Eastern countries.

In the late 1940s new agreements were forged in Venezuela and Kuwait, and in 1952 Iran nationalised all of its oil industry thereby depriving BP of a major source of oil. The weapon was also used again by Nigeria in 1979 when it nationalised all of BP's assets in Nigeria. By 1973 the original concessions in both Venezuela and Kuwait were effectively terminated and companies in these countries operated only under service agreements.

The last major oil province to nationalise its oil industry was Saudi Arabia which was exploited by a consortium of companies known as Aramco, which comprised Exxon, Chevron, Mobil and Texaco. In 1950 the venture held 50% of the concession (and hence 50% of the reserves in the ground) but by 1974 this had fallen to 40%. However, with the rising tide of nationalisation, the Saudi government made it clear in 1974 that it wanted to own 100% of the Aramco concession. After much discussion agreement was reached in 1976 whereby the Saudi government would own the concession and all of Aramco's assets, but Aramco could continue to be the operator and provide technical services to Saudi Arabia for which it would receive $0.21/bbl. In addition, it would also market 80% of all of Saudi Arabia's production. The Saudi government also agreed to pay compensation for all of Aramco's holdings in the country. This deal effectively marked the end of the era when oil companies owned concessions – and hence significant volumes of reserves – in the big producing countries. Today the main producer countries own the bulk of their reserves and the major companies are the marginal producers.

The new state oil companies

In addition to the state oil companies that were formed as a result of the nationalisation of assets, a whole host of new state oil companies sprang up in new oil provinces. This was as a result of the new types of contracts that were being developed in a number of countries, particularly Indonesia. Instead of owning the oil in the ground companies now became 'contractors' with the right to share in the production of any stream of oil that they discovered. These contracts effectively meant that the oil in the ground was owned by the host country. The production sharing contracts also initiated the formation of a number of national oil companies (e.g. Sonotrach in Algeria, Petronas in Malaysia, Pertamina in Indonesia, etc) that were able to 'back into' any venture if exploration activities were successful.

Elsewhere in the 1970s there was real concern that host governments were not reaping the full benefits of the sharply rising oil price. It was argued that major international companies were able to move funds around too quickly and also shelter tax too efficiently. In Britain in the mid-1970s the government of the time decided that it wanted a greater share of the North Sea wealth that was about to flow ashore. As a result in 1975 it initiated a special tax on oil revenues and more importantly established a new state oil company, the British National Oil Company (BNOC). Like other state oil companies that were being formed at the time, e.g. Statoil in Norway, it had the right to buy 51% of North Sea production and was able to buy into new ventures. Although BNOC was privatised in the early 1980s (to form Britoil) state oil companies still dominate the oil industry today.

The independents

Most of the companies that have been discussed in the previous sections are categorised as major companies. However, the growth of these companies left a niche in the market for a number of smaller players or independents who were able to exploit more efficiently some of the non-core assets that the majors were eager to dispose of. The character of these independents varies from one country to another, but their ultimate survival is very much dependent on their ability to add to their value through exploration success or sustain production through acquisitions.

The US independents

The rapid growth of the US oil industry was dominated by the major companies described earlier. However, as these companies grew and focused more effort overseas it became apparent that many US oil or gas producing provinces were no longer the core of their businesses. In particular many of the older producing fields were reaching a point in their production histories where it was no longer viable for the major companies, with their high corporate overheads, to continue operating. As a result, many older fields or properties were sold on to small companies with low overheads and which could make a profit out of the low rates of production. Rather than being full blown exploration companies these companies sought to add value by stripping out overheads and to maintain production by infill drilling in older fields, a strategy which still continues today.

A small number of the US independents have, however, managed to grow to a size where US production is no longer thought to be sufficient to maintain their business. As a result they have ventured overseas in order to boost their reserves base. Union Texas and Oryx are examples of the more successful US independents.

The UK independents

The growth in the UK independent sector has its origins in the early 1970s when many foreign oil majors joined up with small UK oil companies in order to gain access to the North Sea. Access to exploration rights in the North Sea was, and still is, obtained through a series of Government Licensing Rounds in which acreage in the North Sea is awarded to companies on the basis of certain levels of exploration commitment, e.g. seismic surveys, exploration wells, etc. Concerned that British industry would not gain a fair share of the North Sea resources the Government made it clear that foreign oil companies were more likely to gain licences if they combined in joint ventures with British companies. The fourth Licensing Round in 1971 was to prove the most productive of all the rounds, and many small independent UK oil companies were set up as a result. Some of these companies are still in existence today, e.g. Clyde, LASMO, etc.

With the tremendous rates of exploration success in the North Sea throughout the 1970s, there emerged a real appetite in the investment community to participate in that success. Not only did a whole host of UK independents spring up around this time, but many financial institutions and large companies, e.g. ICI, Thomson Newspapers, etc, also took stakes in North Sea assets. Exploration success rates in the North Sea slowed down in the 1980s, and with the fall in the price of oil in 1986 it was clear that the UK independent sector was likely to be rationalised.

This occurred between 1987 and 1990, in which time many oil companies were taken over, as their competitors realised that it was often cheaper to acquire North Sea reserves through take-overs than through exploration. By 1990 the rationalisation of the sector was almost complete and only a few of the original firms survived.

Today the more successful UK independents, like their US rivals, realise that the only way to achieve real growth is through exploration overseas. Companies like LASMO and Enterprise are therefore branching their activities overseas in the hope of discovering larger volumes of reserves than are likely to be found in the North Sea today. Indeed, even some of the smaller companies such as Clyde and Premier have adopted worldwide exploration strategies. However, there is still a niche in the North Sea for smaller firms to pick up pieces of acreage which are non-core to major firms, and make discoveries which may result in a material difference to the size of such a firm, e.g. Monument. However, the days of large exploration discoveries in the North Sea are now over, and unless shareholders are happy with firms simply maintaining production through farm-ins and small discoveries, a further round of rationalisation is likely to occur in the near future.

RESERVES AND PRODUCTION

Oil

Recently published data suggests that the world's total proven oil reserves are nearly 1000 billion barrels. About three quarters of this is estimated to be in OPEC countries, and in particular those in the Middle East. Eastern Europe, China and the republics of the former USSR account for 8% of the world total. The rest of the world therefore has only about 16% of the reserves, and finding sufficient new reserves to maintain non-OPEC production levels will require a sustained high level of exploration activity (see Fig. 1.4).

The continued importance of OPEC in the world oil industry today becomes all too apparent when one considers that over 76% of all proven oil reserves are located in OPEC states. In addition, with 66% of proven reserves located in the Middle East it is clear that in time the oil industry will gravitate back towards that area as reserves are depleted elsewhere. The unsurprising disparity in the ratio of total oil reserves to production shows that the growth in real production will come in the future from the Middle East. With OPEC only producing a small

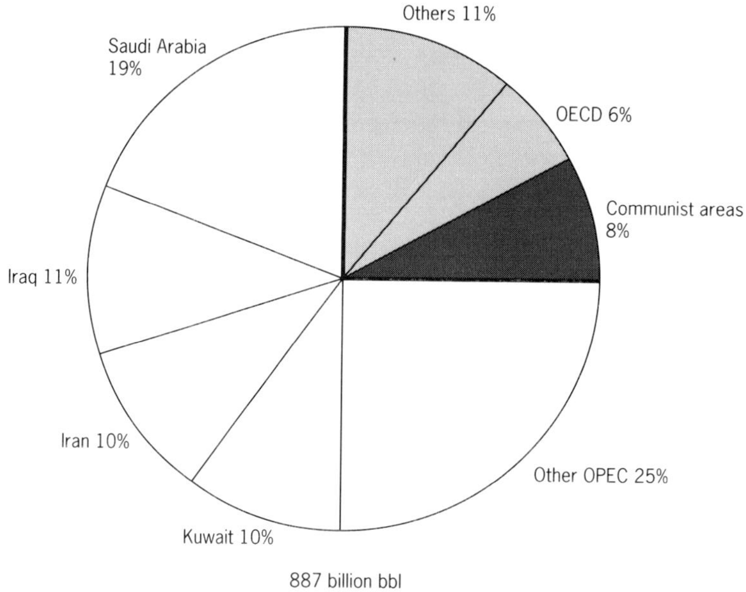

1.4 *World proven oil reserves (source: Holmes, 1988).*

proportion of its own reserves base each year (39% of total world production) the rest of the world will face a difficult struggle in trying to maintain its share of production as reserves are depleted. OPEC's reserves to production (R/P) ratio is (at the time of writing) 87 years, compared with the world total of 46 years (see Fig. 1.5). The higher the R/P ratio the greater is the potential for development. The Middle East leads regionally with an R/P ratio of 104 years and Saudi Arabia's R/P ratio is 86 years.

It is interesting to note that even though only 26% of world proven reserves exist outside OPEC states many of these provinces are described as 'geologically mature', e.g. USA, North Sea, etc, where there is little likelihood of further major discoveries being made. In contrast, the exploration effort in many of the Middle Eastern countries has barely even commenced. If further exploration efforts are made in the Middle East there is no doubt that the percentage of proven reserves in OPEC countries is more likely to be in excess of 85% of the world total rather than the 76% today. Even without a significant exploration effort the Middle Eastern countries managed to boost their reserves base in 1987 through a series of revisions to recoverable reserves estimates. It is likely that such revisions will continue in the future (see Fig. 1.6).

Of the OPEC producers Saudi Arabia has by far the largest percentage

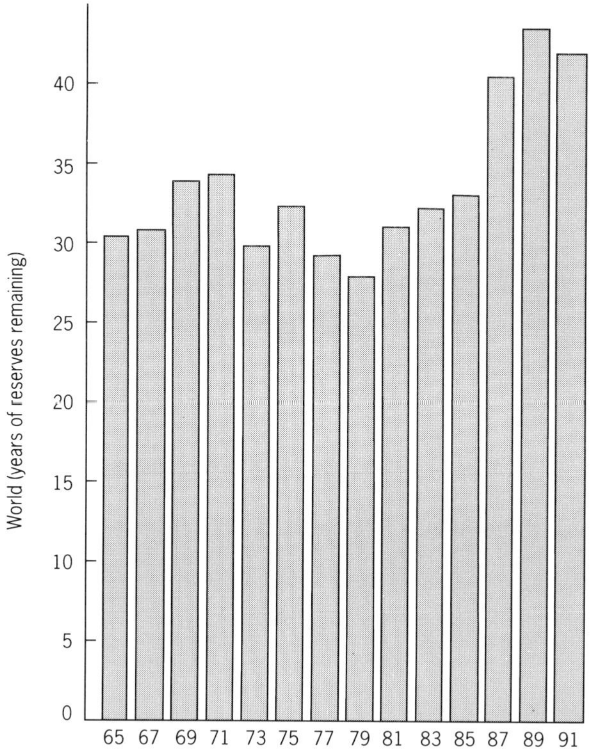

1.5 *World reserves to production rates (source:* Statistical Review of World Energy, *British Petroleum, 1992).*

of oil reserves. In fact Saudi Arabia owns nearly 20% of the world's proven reserves with Iraq, Iran and Kuwait between them owning around 30%. In the light of the fact that none of the concessions in these countries are owned by international oil companies, the reason why the Western world becomes concerned when there is any political turmoil in the area, such as the Iran-Iraq war and the Iraqi invasion of Kuwait, becomes apparent.

Away from the Middle East the largest oil provinces lie in South America, with Venezuela and Mexico dominating. Together both these countries own approximately 11% of the world proven reserves and again both of these countries are dominated by state owned oil companies. Reserves revisions and new discoveries in Venezuela in 1992 resulted in a gain of 3.6 billion barrels which was more than half the world's gain in reserves that year. The country's reserves base has increased steadily over the past few years as Petroleos de Venezuela (PDV) develops huge heavy oil reserves and adds to a string of light oil dis-

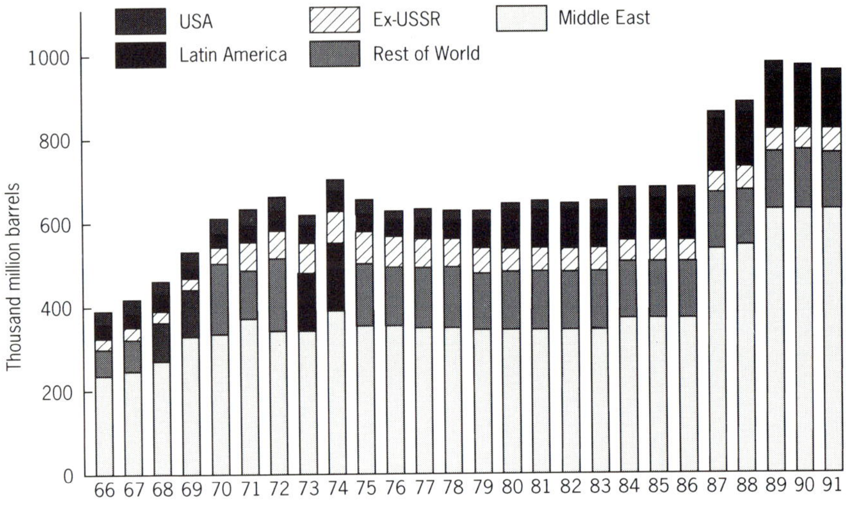

1.6 *The increase in world proven oil reserves (source:* Statistical Review of World Energy, *British Petroleum, 1992).*

coveries in the east of the country. Latin American production now accounts for around 12% of total world production and this proportion has increased steadily over the past twenty years.

After dominating the world oil industry for so long the US now has only a relatively small share of proven resources (3%) even though it accounts for nearly 17% of total world production. With such a large imbalance between its current production and the size of its reserves base it is clear that the importance of the US as a major producer of oil will decline over the next decade. In fact the R/P ratio for the US and Canada is now only 9 to 10 years. Despite the importance of the North Sea to European countries the reserves base is less than 2% of the world total with Norway leading the field. Norway is also the premier producer in the North Sea and is likely to remain so for at least a few more years as total development spending starts to exceed that of the more mature UK sector. The reserves in the UK sector of the North Sea account for only 0.4% of the world's total. This puts into context the activities of the UK independent companies and shows why the more successful participants are actively seeking to expand their reserves bases outside Europe. Like the US, the North Sea is producing a disproportionate volume of reserves each year and in 1991 accounted for nearly 7% of total world production. The R/P ratio for the North Sea is currently only 6 years.

Elsewhere, important regions include Africa (with 6% of total world reserves) with Nigeria being the most important oil producer. Nigeria has perhaps one of the most aggressive exploration and development programmes of all the African states and is planning sharp increases in production from its reserves base which is currently some 2% of the world total. Despite owning some 6% of the total reserves base the CIS shows little sign of increasing in importance until there is political stability and the right environment for investment. With the economies of the former Soviet Republics in turmoil the production decline shows little sign of being turned around before 1995. In 1992 CIS production was down by 13% on the 1991 total, but it still accounted for some 17% of total production.

Gas and liquefied natural gas

Natural gas currently provides some 21% of the world's primary energy demand compared with 38% for coal (see Fig. 1.7). The current proven estimates for gas are estimated to be some 123 000 billion cubic metres (bcm) which is some 60 times the current annual rate of production. Of the total volume of proven gas reserves, over half is located in the former

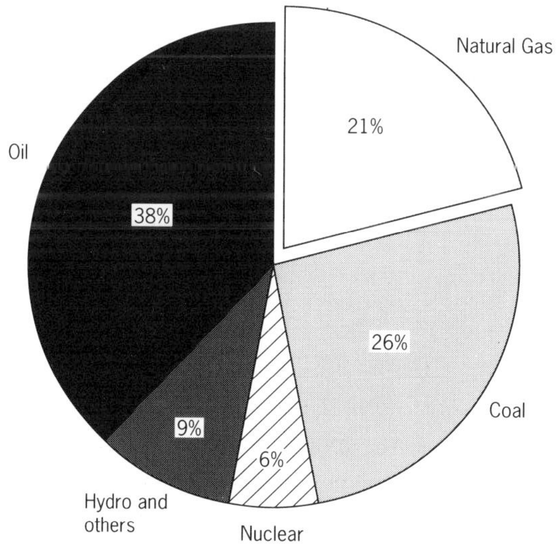

1.7 *World primary energy consumption, 1991 (source: Peebles, 1992).*

Soviet Union and Iran; indeed, over 40% of total reserves are located in the former Soviet Union alone (see Fig. 1.8).

The importance of gas as a primary fuel has grown rapidly over the past 25 years as vast new quantities of reserves have been proven through exploration and appraisal drilling. Discovery rates for gas over the past 20 years have been much higher than those for oil (see Fig. 1.9); over the past 25 years the gas reserves base has grown six times in contrast to the oil reserves base which has grown only two and a half times (much of this due to OPEC revisions). The real growth in the gas reserves base has been in the former Soviet Union, although the Middle East gas reserves base has grown eight times over the past 25 years.

In recent years the average annual rate of discovery of new gas reserves has been four times greater than the annual rate of production. However, the bulk of these discoveries has been in locations which are remote from the main consuming areas. This is an important consideration, as the cost of transporting gas through a pipeline or as LNG is much greater than for oil or coal by sea.

The high cost of transporting gas explains why over 85% of the natural gas consumed in 1991 was locally produced with only 15% being traded internationally (see Fig. 1.10). It is not surprising that with 40% of the

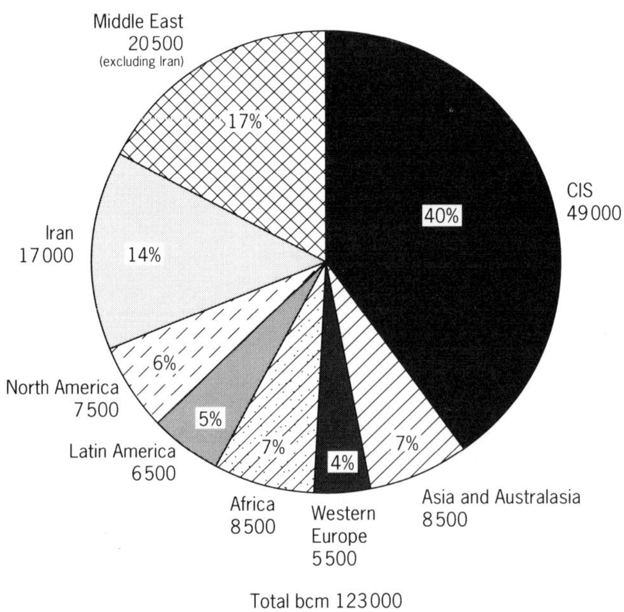

1.8 *World proven gas reserves, 1991 (source: Peebles, 1992).*

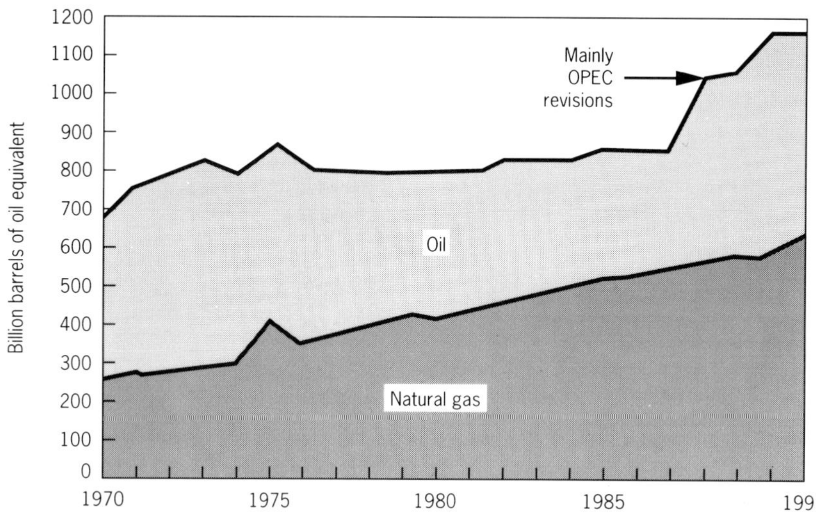

1.9 *Evaluation of proven world oil and gas reserves (source:* Statistical Review of World Energy, *British Petroleum, 1992).*

world's total gas reserves the former Soviet Union should consume some 32% of the world's gas production in 1991 (see Fig. 1.11). However, the importance of the local market is highlighted by the US, which consumed nearly 28% of 1991 total production even though it owns less than 4% of the total reserves base. The importance of the local market

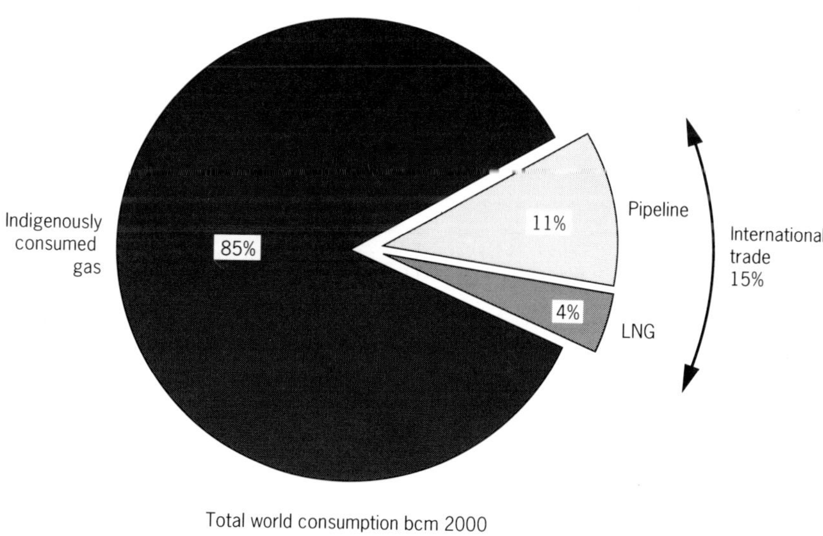

1.10 *International gas trade, 1991 (source: Peebles, 1992).*

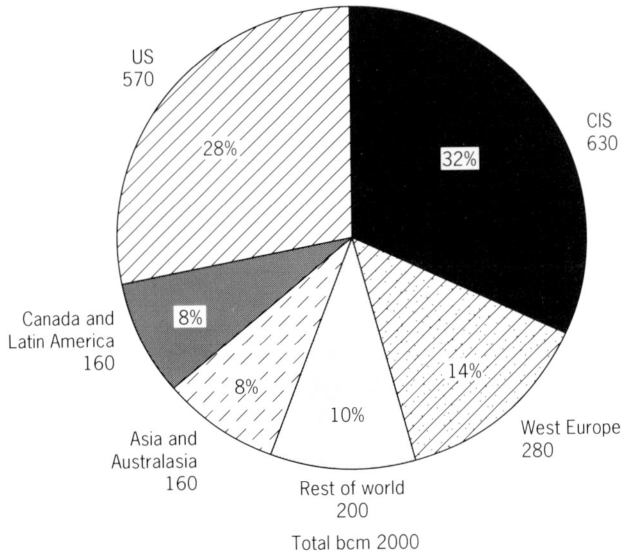

1.11 *World gas consumption, 1991* (*source:* Peebles, 1992).

has also been highlighted recently in the UK where the lack of a coherent energy policy has meant that new gas projects may not be approved. The impact of cutting off the local market obviously means that many projects will have to be deferred.

The real problem for countries with large reserves of gas is how to market it internationally at competitive prices. Of the 15% of production which is being traded internationally nearly 75% is transported by pipeline. The CIS is the largest exporter of gas with over 106 bcm being bought by Western European countries, primarily Germany. Canada is the second largest exporter of gas and has a growing market in the US. In Europe the primary supplier of gas is the Netherlands which has Germany as its primary market. Algeria is also growing in importance as a gas supplier and has a ready market in Italy.

For countries from which it is physically impossible to export gas reserves through a pipeline, e.g. Indonesia, the only way to transport the gas is as LNG. Although LNG's percentage contribution is relatively small, for Japan and South Korea it is for all intents and purposes the only source of gas supply. For some other countries, e.g. France and Spain, it is a significant component of their total gas supplies. Italy, Taiwan and the US also import LNG and with other countries looking to import in the near future, the role of LNG looks set to increase in importance in the next decade. Currently the biggest supplier of LNG is Indonesia which exports most of its product to Japan and South Korea. Algeria is also an important supplier of LNG to France but looks set to break into the growing Far Eastern market in the near future. Other major suppliers of LNG include Malaysia and Brunei. With several developments in progress in the Middle East (Oman and Qatar) this area could also become an important exporter in the next decade.

FUTURE POTENTIAL

Every oil company that is committed to the upstream oil and gas business must aim at least to replace the oil and gas it produces each year. It must do this through a judicious balance of finding, at as low a cost as possible, new reserves, upgrading existing reserves, and buying new properties. As discussed earlier, the opportunities for finding huge new fields are now quite slim, although not impossible as BP has recently proved with its Cusiana discovery in Colombia. Even in the mature areas there are still plenty of opportunities. For example, the oil industry in the US expects to find at least as much oil in the next 20 years as it has done in the past 20. Even in the relatively mature UK sector of the North Sea the industry expects at least 2 billion barrels of oil to be discovered in the next decade. Away from these mature provinces there are good possibilities of further discoveries in the world in provinces which, by comparison, are still virtually unexplored.

Industry estimates suggest that between 350 and 430 billion barrels of oil remain to be discovered in the world (see Fig. 1.12). Of this total, some 95 to 120 billion barrels are expected to be discovered in the Middle East where the exploration effort to date has been relatively small. Some 75 to 80 billion barrels are expected to be discovered in the former Soviet Union although it may be some time before any of this potential is realised. In the mature province of North America 55 to 75 billion barrels are expected to be discovered; the bulk of these new reserves will effectively be proved up from exploratory appraisal drilling. However, new geological plays in the deep waters of the Gulf of Mexico as well as the possible opening up of Alaskan acreage could hold sizeable new discoveries. In Europe, at least 20 billion barrels are forecast to be discovered. With all of these hydrocarbons waiting to be discovered

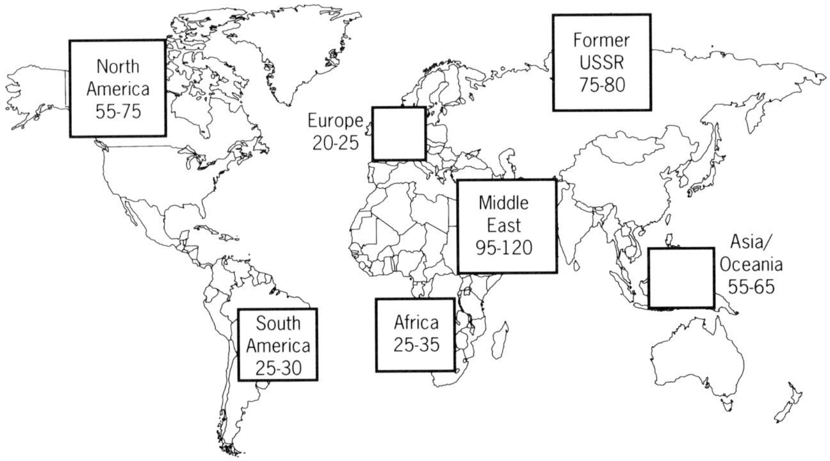

1.12 *Potential new oil discoveries (source: Holmes, 1988).*

there is no doubt that the oil industry will remain active for some time to come.

Even without new discoveries the oil industry could continue to produce at 1992 rates for another 46 years on proven reserves alone. If the future reserves revisions and new discoveries are taken into account this would extend the reserves' life by some 20 years. In addition, there is also the potential for Enhanced Oil Recovery (EOR) projects which could extend production life by as much as another 75 years and this excludes the recoverable reserves from bitumen and shale oil.

Exploratory status of the world's oil provinces

Of the world's prospective sedimentary basins only 27% currently produce hydrocarbons. Another 40% of these basins which have been partially and moderately explored and tested have not yielded commercial quantities of petroleum. The remaining 33% of the world's basins have witnessed little or no exploration activity.

Of the world's sedimentary basins approximately 30% lie in the oceans. These ocean basins include some of the most productive offshore areas such as the Gulf of Mexico, the North Sea, Persian Gulf and Lake Maracaibo. However, there also remain many essentially unexplored ocean basins such as the Bering Sea and the northern Norwegian Sea which have high potential for future petroleum discoveries.

The remaining 70% of the sedimentary basins lie onshore. However, the large areas of onshore basins do not necessarily correlate with large volumes of hydrocarbons. For example the Middle East has only a relatively small prospective area yet it contains 66% of the total world's proven reserves. Onshore basins hold high potential in many parts of the world but the majority of the more prospective basins are in frontier areas, located in harsh physical environments, such as the Arctic, and in the deep water or the remote interiors of continents where any new discoveries are unlikely to be currently commercial.

The productive basins of the world have been divided into two categories: mature and developing. Many explorationists believe that as much as 30% of the undiscovered oil and gas lies in today's mature producing areas. Many of the Middle Eastern basins have not been fully evaluated with recent discoveries being made in geological formations that were previously thought to be unprospective. The potential petroleum basins of the world are essentially unexplored. However, first steps to evaluate the potential of these basins are being made around offshore Argentina and Brazil, and off the east coasts of India, Vietnam and China.

THE OIL INDUSTRY TODAY

The current structure of the oil industry is dominated by the state oil companies which have continued to rise in prominence over the past 20 years. The *Petroleum Intelligence Weekly* survey of world oil companies, which takes into account 6 measures of size and throughput, suggests that of the top 20 oil companies today 11 are state oil firms of which 4 lie in the top 6 (see Fig. 1.13). The largest state firms are Saudi Aramco and Venezuela's PDV. Close behind these two firms are Mexico's state company Pemex and the National Iranian Oil Company (NIOC). The two largest major private oil companies are Royal Dutch/Shell and Exxon. In general terms over the past 20 years, the state owned oil companies have replaced the private firms in the ranking as governments have nationalised their industries. Therefore the private oil companies sector in which the investor is likely to be interested is very small in proportion to the size of the oil and gas industry as a whole.

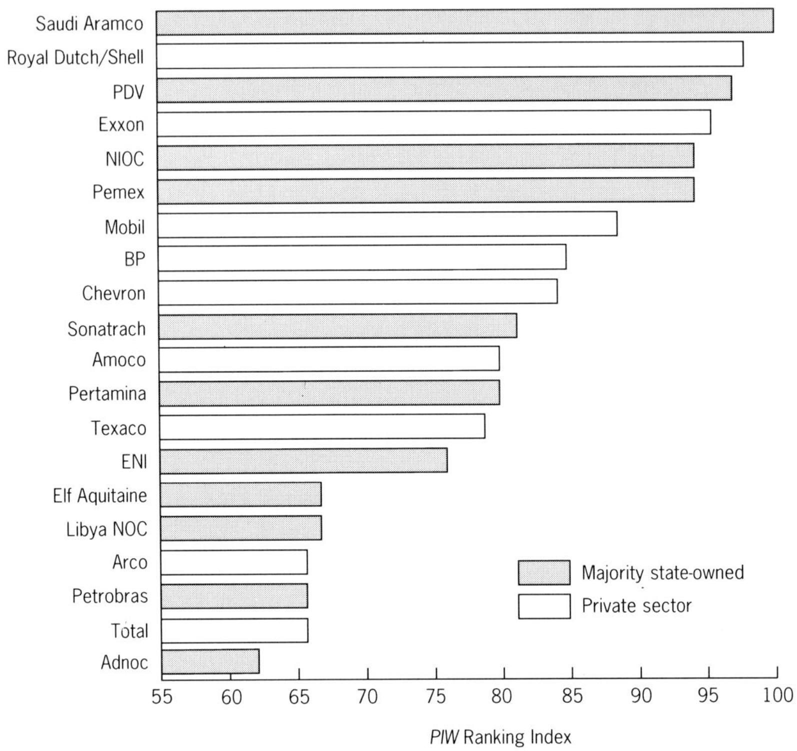

1.13 PIW *ranking of top 20 oil companies (source:* Petroleum Intelligence Weekly, *1992).*

In terms of the ownership of reserves there is no doubt that nationalisation and the trend towards service contracts have left the private firms with only a very small share of the actual reserves in the ground (see Fig. 1.14). Over 92% of all currently proved oil reserves are owned by the state companies leaving only 8% for the private companies. In addition, nearly 90% of all proven gas reserves are owned by the state firms. Therefore the private oil companies which the investor and analyst are covering are in fact the owners of only a very small proportion of the world's known reserves, a fact that should not be forgotten when considering the sector. Indeed additional political risks should be considered before investing in private companies which operate in areas where the political regime is unstable and where reserves could be nationalised after considerable investments have been made.

The dominance of service contracts and production sharing contracts mean that private oil companies fare a little better in their share of world oil production. In 1991, private oil firms owned just over 24% of total

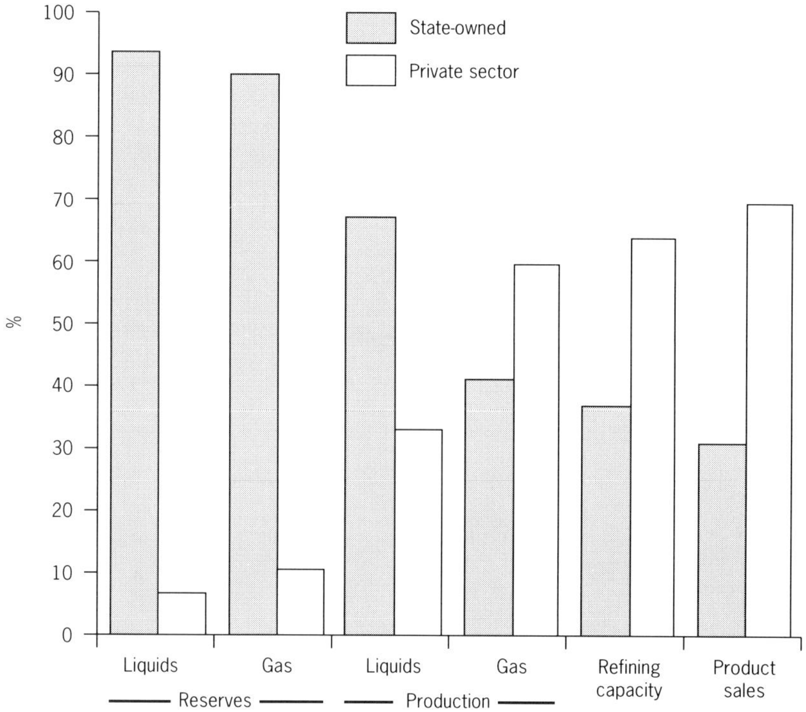

1.14 *Ownership of proven gas reserves (source:* Petroleum Intelligence Weekly, *1992).*

world oil production, and an even greater share in gas (over 50%). The world's top 15 oil producers accounted for over 50% of the world's output in 1991, and this group of companies is dominated by state oil companies with only Royal Dutch/Shell in the top 10. The international majors still dominate the natural gas business. Shell and Exxon top the list of gas producers, with Algeria's Sonotrach third, and Amoco and Mobil completing the top five.

The further 'downstream' one looks into the business, the more the international private oil companies can be seen to dominate. Private firms own over 50% of current refining capacity, and over 60% of all product sales. However, even in the downstream business the state companies are beginning to encroach upon the position of the majors, and companies such as PDV, Aramco and NIOC have all risen in the ranking over the past few years. This trend is likely to continue as state oil companies begin to take the lead in downstream activities, having already consolidated their upstream position.

For the investor looking for sustainable growth potential in the future it should be remembered that the bulk of conventional new oil discoveries are likely to be made by the larger companies which have the greater financial and technical resources. As a result, the larger companies are more likely to gain a greater share of these resources than the smaller independents, which will continue to carve out a niche in market conditions which will be specific to different areas of the world.

CHAPTER TWO

*E*conomic assumptions

A NON-RENEWABLE RESOURCE

Club of Rome

A little less than 20 years ago there was a clear consensus on the workings of the world oil market. The most influential economics document of the time, the 'Club of Rome' report, encouraged an obsession with the need to conserve non-renewable resources. Oil, like minerals, was a non-renewable natural resource with, in the early 1970s, proved reserves which were only equivalent to some 30 to 35 years of world consumption. Demand, it was believed, was fairly impervious to movements in price, and followed growth in economic activity. Future supply capacity was constrained by the fact, it was again believed, that most of the world's potentially prolific basins had already been explored, and the big potential finds made. On top of this gloomily Malthusian analysis, which was then applied to all non-renewable resources, there was an additional threat to oil prices. The largest part of the world's oil reserves had, since the great discoveries in Saudi Arabia in the 1950s, been in the Middle East.

With the formation of the Organisation of Petroleum Exporting Countries (OPEC) in 1960, the Middle Eastern countries, and certain other members, notably Venezuela, had become vociferous in their claims that they were receiving an unduly small share of the economic rent associated with their production. The events which ensued have been covered in the previous chapter of this book, as has its aftermath, the collapse of the 1980s, and the change in OPEC strategy.

Flat commodity prices

Alongside this abrupt change of oil market behaviour there was an equally dramatic change of perception among economists and oil market forecasters. Oil was not the only commodity price to fall during the 1980s. Indeed, it was a decade in which the price of most commodities declined, at least in real terms. Accordingly, a new set of axioms became fashionable. As economies become more sophisticated, it was argued, the relative value of basic commodities as against finished products should tend to decline. The Ricardian assumption that the cheapest reserves will be the first to be extracted, implying rising extraction costs as these are depleted and replaced by higher cost sources of supply, would be offset, it was now assumed, by the cost reduction associated with new and more efficient technology. The very long term real oil price history which was shown in the previous chapter certainly appears to support this view. (Ricardo was primarily concerned with the price of agricultural produce, but he was one of the first economists to realise that commodity prices are set by the cost of the marginal producer, and that the advantage derived from lower cost producers can be analysed as a form of rent.)

In Saudi Arabia, the dominant member of OPEC, it came to be believed that the way to maximise long term revenue was to support a price which would leave oil competitive against other fuels (substitution had been the main cause of long term volume loss in the early 1980s), and discourage the continuing allocation of resources to relatively high cost non-OPEC areas of production. More than this, it was believed, was simply unsustainable, and the volatility implied would only damage demand.

Assumptions for evaluation

The requirement for an evaluation of oil companies and their assets is to obtain a methodology for establishing reasonable assumptions for the crude oil price and for refining margins which can be applied to the subjects of analysis. The remainder of this chapter draws on historical events and data only to the extent required to indicate the factors which should be taken into account when deciding on such assumptions.

It is inevitable that these, like most macroeconomic forecasts, have an extremely short useful life, which is why the prices of shares in oil companies, and the prices paid by oil companies when they buy assets from one another, fluctuate widely over time. Evaluations are only as good as the assumptions on which they are based, and, while it is inevitable that these will be subject to change, they might as well be as objectively based as possible. The chart produced by Shell (Fig. 2.1) of oil price forecasts and subsequent events illustrates the point clearly.

In reality, oil companies (though not, it often seems, investors) make their decisions not only on a base case but also screen their investments on a worst case scenario, and reject decisions which are uneconomic on

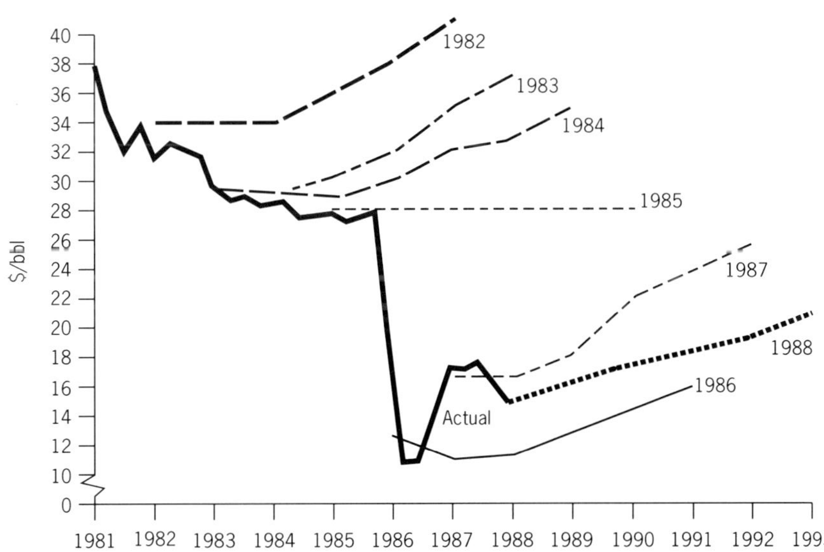

2.1 *Typical industry crude price views (money of the day) for Arabian Light or equivalent (source: Shell).*

this basis. Particularly in the case of integrated oil companies, there are a wide variety of external variables which drive their profitability. For their oil and gas operations, three predominate, namely the dollar crude oil price, and, in their individual geographical regions of operation, the dollar refining margin per barrel and the wellhead gas price. We shall concentrate on these in most of the remainder of this chapter.

Analytical issues

With some oversimplification, it could be argued that there are three main complications to oil market analysis:

1 The first, on the demand side, is that the demand for crude oil is a residual demand, composed of the demand for the range of products into which crude oil is refined. The sensitivity of world demand for each of these products to fluctuations in economic activity and to price is different.

2 The second, on the supply side, is that although OPEC reserves are mainly low cost, OPEC acts as a residual supplier, attempting to optimise its revenues by setting its supply in the light of the relationship between the volume of demand for its oil after non-OPEC has produced to full capacity and the price which it will achieve for a given level of production.

3 The third is that, both for demand and for non-OPEC supply, the sensitivity of volumes to price is much greater in the long term than in the short term. We shall look at each of these points individually, and attempt to pull them together into a simple model of how the crude oil market works.

OIL MARKET DEMAND

Residual demand

Product types

Although crude oil is refined into a wide variety of finished products and feed stocks, for economic purposes its uses fall into three main groups:

transport fuels, heating oils and fuel oils. These categories, by end use market, are the most appropriate to the economist attempting to forecast oil demand. Unfortunately, they do not correspond exactly with what happens in a refinery (covered in some detail in Chapter 5).

For the historical data reproduced in this section, we have used BP's 'Statistical review of world energy (June 1992)', a source which is both compendious and clearly presented. As with many other sources, consumption figures are given by category of oil products, namely gasolines, middle distillates, fuel oil and others:

- The gasolines category consists of light transport fuels, aviation and motor gasolines, but also includes light distillate feed stocks (for petrochemicals plants, for example).
- The middle distillates category includes kerosene (most of which is used for jet fuel but which is also used for heating, especially in Japan), diesel oil (most of which is used as automotive and marine transport fuels, but which is used in engines for other applications) and gas oil (for which the largest use is domestic heating, but which also has uses as industrial fuel).
- The fuel oil category includes fuel for marine bunkers, as well as onshore uses as industrial fuel and for power generation.
- Finally, the 'other' category contains a diverse range of products and applications, from bitumen to lubricating oils.

Product demand curves

Clearly, to examine in detail the effect of price and economic factors on oil product demand, it would be necessary to delve into the history of each individual application of the main petroleum products, and to differentiate between the regional behaviour of the markets. Most econometric models of oil demand fail to do this.

For any given point in time and level of economic activity, the demand curve for oil can be conceptualised as beginning with the call on oil products for which the cost of substitution (or significant conservation) is such that the amounts consumed would be fairly insensitive at higher prices. It continues with demand for products for which substitutes can compete, or for which significant efficiency savings can be achieved at a price. It ends with demand for products (fuel oil for power stations, for example) by consumers which can virtually costlessly switch between different fuel sources. In economic jargon, the price elasticity of the demand for aviation kerosene is massively lower than the price elasticity of demand for domestic heating oil, because it is almost impossible to substitute for aviation kerosene.

There are step jumps in this series; it is not continuous. For quantitative work, it is also important to note the importance of taxation to the price paid by the customer for petroleum products bought on inland markets. In the case of gasoline, for example, European excise taxes can represent the larger part of the cost to the motorist, and this no doubt explains the apparent insensitivity of demand for this product to price. A large percentage increase in the crude oil price, which will probably be reflected fully in the wholesale gasoline market, may only result in a fairly small percentage increase in the retail price to the customer.

Tax effects

The extent to which the fall in crude oil prices has been absorbed by European governments, rather than being allowed to pass through to the customer, is illustrated in the chart showing indexed European consumer prices for oil products and the OPEC crude oil basket price from 1985 to 1992 (Fig. 2.2).

Demand by product

Introduction

The charts of non-Communist oil product consumption by geographical region illustrate two factors: namely, the changing trends in consumption by product group in each region, and the different trends in overall consumption by region for the period since 1965 (see Fig. 2.3). Taking the trends by product first, it is evident that the main long term effect of the second oil crisis, of 1979-81, was to reduce considerably the consumption of fuel oil in the North American and Organization for Economic Cooperation and Development (OECD) European markets. This is intuitively where one would have expected to find the greatest price effect.

Fuel oil

Fuel oil is used in the provision of thermal energy and in the generation of electricity, both of which are amongst the more easily substitutable uses of oil. In the political mood which followed the oil crisis, it proved

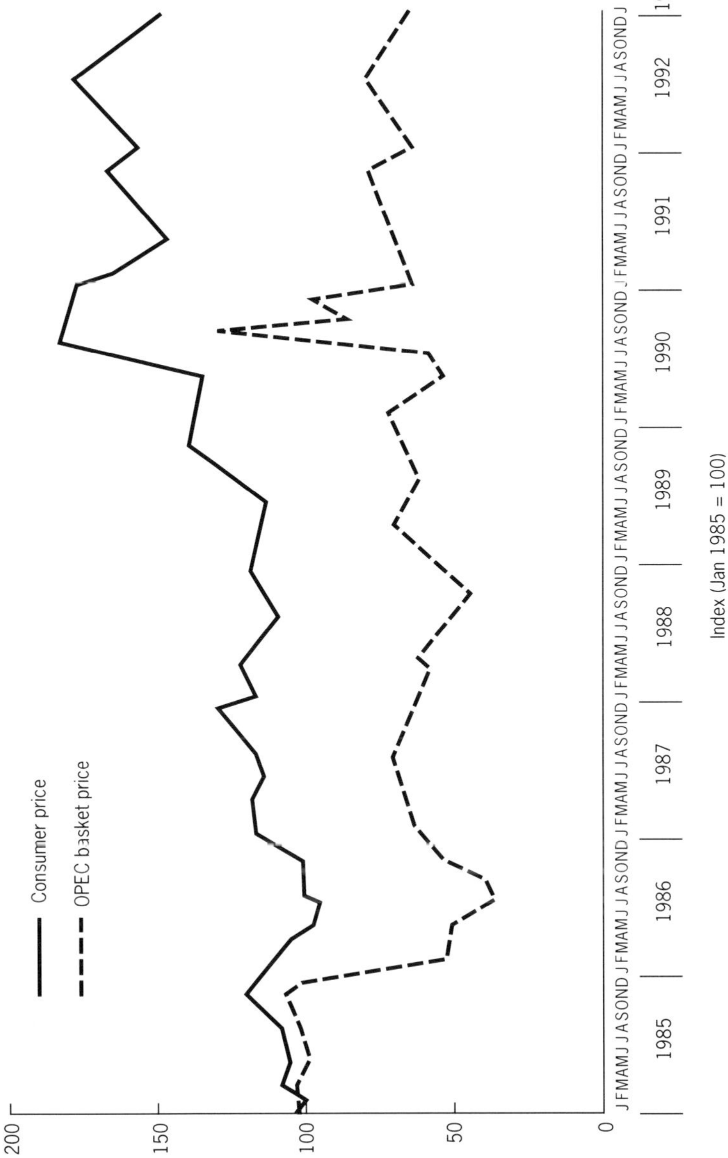

2.2 *European consumer prices for oil products compared to the OPEC crude oil basket price (source: Petroleos de Venezuela).*

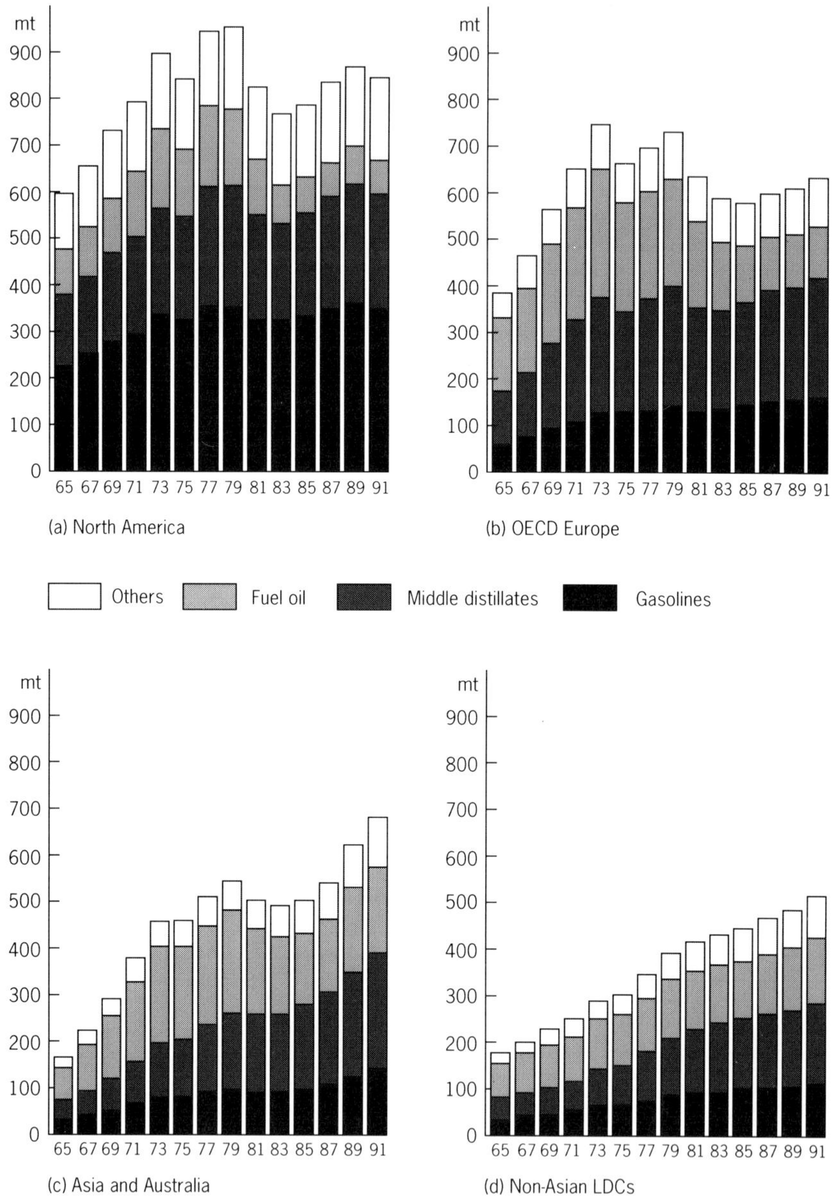

2.3 *Oil consumption by product group in North America (a), OECD Europe (b), Asia and Australasia (c) and non-Asian LDCs (d). (Source:* Statistical Review of World Energy, *British Petroleum, 1992.)*

fairly easy for most of the Western OECD countries to achieve support for expansion of nuclear generation programmes, and this combined with increased use of alternative fossil fuels (coal and natural gas) to result in considerable substitution out of oil in these areas. However, with environmental regulations pressuring coal use, there may be some short term switching back to fuel oil in the 1990s.

The scope for further reduction in North America is now limited, since much of its fuel oil consumption is now committed to marine bunkers, where substitution is more difficult. In Europe, the situation differs by country. In Italy, for example, oil is still heavily used for power generation, and is likely to be displaced by natural gas from North Africa. The effect of higher fuel oil prices on fuel oil consumption was not as great in Asia and Australasia; there was an initial decline in the early 1980s, but this was all subsequently recovered. These are mainly faster growing economies, and more importantly, they have either technical or political inhibitions against nuclear power, and have not had a complex of gas pipelines in place to facilitate substitution.

The same argument applies in an extreme way to the less developed countries (LDCs) of what used to be called the non-Communist world. These have for the most part enjoyed rapid economic expansion, and are at the stage of development where high energy intensity is a feature of their growth. They lack the capital and the technology to switch to alternative fuels, and their consumption of fuel oil has thus continued to rise.

Gasoline

At the opposite extreme is the market for gasolines. Although depressed by the recessions of 1975 and 1981-1982, gasoline consumption in all markets rebounded strongly during the 1980s with the expansion of passenger miles travelled. This was partly, but not fully, offset by increased automotive efficiency, which absorbs perhaps 1% annually of potential increase in demand.

Middle distillates

The pattern for middle distillates is similar to that for gasoline, but has been subject to some substitution effects, again mainly in the US and European markets in those areas where substitution is possible, notably heating. The prospects here are less clear. Both in the US and in parts of Europe, e.g. Germany, there remains considerable scope for substitution of gas oil by natural gas for heating purposes, though whether this will occur or not is dependent on whether large volumes of additional natural gas can be delivered to these markets at prices which are competitive with oil products. But the US gas market has been damaged by clumsy regulation and European gas prices are rising, so significant further substitution may not occur unless oil prices rise sharply.

The most rapid growth of all has been in Asian and LDC consumption of middle distillate, and this seems set to continue for some years absent a major price shock. Given the composition of products which they represent, it is unsurprising that there is a fairly stable trend to the demand for the category of other products. These tend to be either high value added, and therefore fairly price insensitive, or related to the construction industry.

Demand by region

History by region

Slicing the figures the other way, by region rather than by product, it is obvious that the maturity displayed by the North American and OECD European markets is not at all matched in the Asian and other LDC economies. The Asian and Australasian group saw very rapid demand growth in the late 1980s, such that the setback which followed the 1979–1981 oil crisis was completely reversed. Remarkably, demand growth in the non-Asian LDCs was completely impervious to oil price shocks.

Changing composition

This has had important results for the changing composition of international oil consumption, and therefore for the likely behaviour of demand in future. In 1979, North American oil consumption was some 950 million tonnes, against about 540 million tonnes in Asia and Australasia, and about 390 million tonnes in the LDCs outside these areas. In 1989, prior to the recession or the Gulf crisis, North American consumption of 850 million tonnes was dwarfed by figures of 610 million tonnes from Asia and Australasia, and 500 million tonnes from the non-Asian LDCs.

LDC evolution

The time will probably come when greater per capita incomes, greater capital resources, access to more sophisticated technology and an established infrastructure will alter the demand characteristics of these economies into something closer to the US or OECD European model. It is probably some years away, however, and this suggests that, in the absence of draconian environmental legislation, demand for oil products from the old non-Communist world is likely to be quite robust against oil price fluctuations, even if it continues to lag economic growth rates somewhat.

Ex-Communist countries

Integration into markets

Following the collapse of Communism in Eastern Europe, it has become less appropriate to model the world oil economy by excluding these economies except to the extent of their exports to the West. They are all, including the former USSR republics, raising oil prices to world levels, and moving away from central planning with its emphasis on fixed export volumes.

The data problem

In contrast to the theoretical arguments for inclusion of the former Communist countries, there is still a great practical difficulty in obtaining data about consumption in these economies, and most estimates now use derived demand figures (simply subtracting exports from production to impute consumption). Moreover, price liberalisation means that historical data is of less use than usual when forecasting the future.

China

As far as the oil industry was concerned, there was always a large difference between China and the Soviet Bloc. China has behaved much more like non-Communist producers, in permitting equity participation by oil companies (though the high hopes of the early 1980s for the potential of the South China Seas were sadly disappointed). And in terms of its propensity to consume, China, or rather the coastal provinces of China, are behaving much like the other industrialising economies of the Far East.

Soviet Bloc

The Soviet Bloc was, and is now, rather different. Traditionally supplied by Russia, oil was priced at about a quarter of world prices. Thus, in addition to the economic contraction associated with the opening up of their markets to the West, these countries have recently undergone an oil price shock similar to that which the West suffered in 1973/4. It may well be, therefore, that even when their economies begin to catch up with those of the West, this will not correspond to large increases in demand for oil.

The same comment, to an even more extreme extent, applies to Russia itself, but Russia is also an enormous oil producer. The economic collapse which preceded and followed the fall from power of the Communist Party accelerated the decline of production from Russian oil fields. The net effect for the first two years was for supply and demand to fall almost in tandem, with only a small decline in exports.

It may well be that by the time Russia's undoubtedly substantial potential production capacity is being efficiently exploited, there will have been a sufficient overall improvement in the general economy that the surplus will only rise gently if at all. This may become one of the more crucial issues for the oil market in the second half of the 1990s, with Russian exports possibly varying by three million barrels per day between high and low estimates.

Environmentalism

Emissions

Environmental legislation is having a significant impact on the oil industry in a variety of ways, most of which simply increase the cost of refining or transport for the oil companies. The ever tightening restrictions in both the USA and Europe, from removal of lead as a gasoline additive to lower sulphur content, to the draconian requirements of reformulated gasoline, or the new requirements for tanks at petrol stations and tanker hull design, all come into this category. To the extent that the companies pass this cost on to their customers, it reduces demand, but the effects have probably been fairly marginal.

The 'greenhouse effect'

Worries over the greenhouse effect come into a completely different category. Whatever the scientific evidence, there is an increasing consensus that carbon emissions, which result from the burning of all fossil fuels, should be constrained, and the first moves are being made on both sides of the Atlantic to achieve this.

The likely impact of these moves is, at time of writing, unquantifiable. To date, the heaviest excise taxes have been levied on gasoline, precisely because it has a relatively insensitive demand, and most governments in the developed world (the US is an exception) have exploited this. The object of general carbon taxes would clearly be the opposite, i.e. to raise the cost of energy in ways which would encourage reduction in its use. Such taxes would also, depending on how they were raised, encourage substitution from those fuels which in use emit large amounts of carbon dioxide per unit of energy produced (notably coal), to those with a lower environmental impact (notably natural gas). So there might be some compensation for the oil industry.

Uncertainty

If we have made little allowance for the effect of environmental legislation on demand, or on the behaviour of OPEC (discussed later), it is because we do not as yet have adequate information to assess its significance, not because it will not be significant.

NON-OPEC SUPPLY

Ricardo reversed

A free market

In a perfect oil market, the cheapest reserves would be tapped first. In the actual oil market, cheap reserves in the Middle East are left undeveloped or, even more surprisingly, developed and then depleted at a rate which is well below their full production capacity, while relatively more expensive oil is produced to full capacity and oil companies explore for oil which would be rejected as uneconomic if the market were to operate in an unregulated fashion.

OPEC intervention

The above paradox is attributable to OPEC price support, a comment which may appear to contradict one made earlier, to the effect that OPEC pricing is now aimed at damping oscillations around a clearing price. In fact, the paradox largely disappears if a distinction is drawn between the short run marginal cost of supply and the long run cost of supply. Arguably, in addition to its discovery that demand for at least some oil products is sensitive to price, OPEC also established, through its pricing policies during the course of the 1980s, a much clearer understanding of what the long run marginal cost of its competitors' supply actually is.

Production cost structure

Chapter 4 of this book, on the evaluation of exploration and production companies, looks in detail at the economics of individual oil and gas fields. A key point is that there are three elements to the cost of producing a barrel of oil, prior to royalties or other forms of taxation, namely, the cost of finding and appraising the field, the cost of developing the field (including the cost of required infrastructure, such as pipelines), and the operating cost of producing incremental barrels once the field has been brought on stream.

The balance between these costs varies massively by region. For onshore wells in the mature parts of the US the capital costs are negligible, but the production costs may be high to reflect the recovery techniques required. In the UK sector of the North Sea the capital costs are typically very high as they include vast platforms and subsea pipes, but the wells mainly produce large volumes per day without expensive stimulation.

Short and long run marginal cost

When discussing the cost of a barrel of oil it is crucial to differentiate between the cost of producing oil from developed fields (equivalent to the short run marginal cost of production) and the cost of finding, developing and producing new fields (equivalent to the long run marginal cost of production).

The US example

The characteristics of US onshore oil are such that one would expect it to respond fairly quickly to fluctuations in the price of oil, as falls can easily make it uneconomic to produce even from developed reserves, while higher oil prices make it worth drilling incremental wells with an almost immediate impact on production. Despite its sensitivity, however, its maturity implies a downward long term trend. In reality, this is exactly what has happened. In price shocks, such as that which followed the Iraqi invasion of Kuwait in the summer of 1990, US production is quick to react upwards, and in the summer of 1986 it was the approaching insolvency of much of the US oil industry (and its banks) which led the then Vice-President Bush to fly to Riyadh in a successful attempt to change Saudi oil policy back away from a free market.

The UK example

The characteristics of UK offshore oil are precisely the opposite of the US example. Once a field is in production, the cost of producing additional barrels is negligible. Moreover, the very high marginal rate of tax has been such that a significant proportion of the loss in cash flow due to lower price has been borne by the Inland Revenue, rather than the oil companies. Despite the tax effect, however, the economic attractions of exploring for and developing new oil fields are impaired long before the continued depletion of currently producing fields comes into question. The long run marginal cost of supply is much higher, in other words, than the short run marginal cost of supply.

Reserves and production

Production history

Figure 2.4 of non-OPEC production by geographical region illustrates the crippling effect which the medium term response of non-OPEC

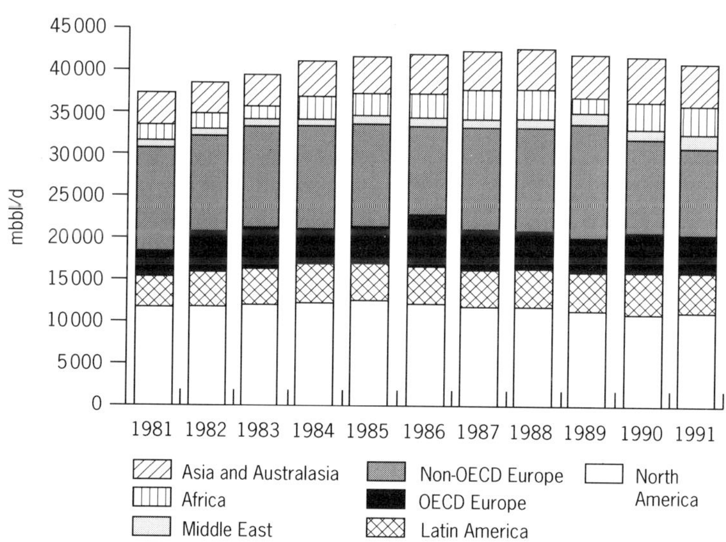

2.4 *Non-OPEC production (source:* Statistical Review of World Energy, *British Petroleum, 1992).*

production to the second oil crisis had on Middle Eastern production between 1979 and 1985. In addition to flagging demand, OPEC was confronted by expansion in output from its competitors – notably in Europe, Asia and the non-OPEC countries of the Middle East.

The price collapse of 1986 then further demonstrated that, outside North America, this additional production, once developed, was insensitive to a fall in the oil price to pre-crisis levels; it was there to stay. But if the high oil prices of the early 1980s had the effect of encouraging the rapid development of oil reserves in non-OPEC producing regions, it had no similar effect on the rate at which these regions added to their reserves by successful exploration.

Reserve history

Illustrating the history of proved oil reserves worldwide, Fig. 1.4 shows that reserves have been maintained, but not greatly increased, since the mid-1970s. The sharp increases in estimated world oil reserves, which had the effect of increasing the ratio of world reserves to annual production to a record of over forty years, were entirely attributable to revisions of estimated reserves in Middle Eastern OPEC countries. While there is little doubt that these figures had been understated during the early 1980s, it is also likely that they were then inflated to the full at a time when it seemed that the allocation of OPEC quotas would begin to reflect reserves.

Future production

There is always a tendency to underestimate future non-OPEC production. Since it is generally operated at close to full capacity utilisation, the production profile of its currently producing fields can only diminish over time, as the fields deplete. The flat world reserve history shows that this tendency has historically been offset by exploration success, which has permitted this decline to be made up for by production from new fields.

Although there may not be obvious new regions for the industry to explore, this may be offset by the effect of political changes in established prolific sources of oil. Owing to the perceived likely shortage of capital in the later 1990s, several OPEC members including Venezuela and some of the Arab countries are opening up to foreign equity investment upstream. As many of these areas are relatively underexplored, there must be scope for significant upward reappraisal of reserves there. The opening to Western oil companies of parts of the former Soviet Union may also add significant new reserves.

Supply curve analysis

Future discoveries
It may be reasonable to assume that the trend detailed above will continue, and that, at stable real prices, future non-OPEC exploration will continue to balance the effect of depletion. Certainly it would be wrong simply to extrapolate the falling production which will be generated by currently known reserves.

Costs best guide
Probably the best approach is again to use costs as a guide. In the UK North Sea, for example, the region has become mature, and the size of exploration finds is much smaller than it used to be. On the other hand, the extent of its infrastructure is such that the per barrel capital cost of bringing new finds into production is much lower, and the success rate of new exploration wells is much higher, owing to improved technology and greater experience of the geology of the area. Thus, the full cost of incremental UK reserves has fallen considerably in real terms over the past decade. In addition, lower cost areas are opening up to the industry, due to political liberalisation.

Supply curve elasticity
Rather than thinking in terms of point forecasts for each year, non-OPEC supply should be seen in terms of a series of supply curves for each

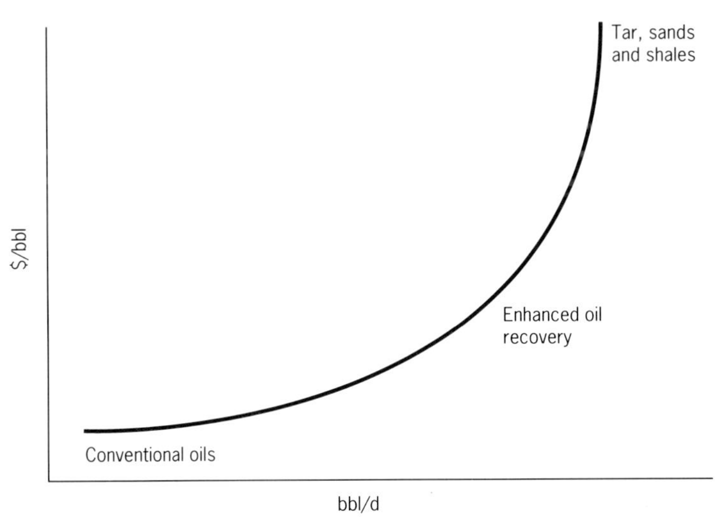

2.5 *Non-OPEC supply curve.*

year, albeit curves that may shift from year to year as the cost structure of available reserves changes (see Fig. 2.5). The broad outline of the range of reserves which make up the non-OPEC supply curve change only slowly over time. At one end is the lowest cost conventional production, in areas such as Oman or Brunei, for example, and at the other end are heavy oils, tar sands and shales, which can add enormously to production but at much higher prices.

Supply curve movement

When forecasters project a flat production profile for non-OPEC, and premise this on flat real oil prices, what they are in fact doing is assuming that the supply curve for oil from these regions will not shift in real terms over time. This may be a reasonable assumption, though intuitively it might seem more likely that in fact there will be a long term tendency for costs to rise in real terms, implying either higher real oil prices or falling non-OPEC production if prices stay flat. Saudi oil policy after 1987 was premised on the opinion that the long run cost to non-OPEC of maintaining its production is in the mid-teens, in 1987 US dollar terms (which should be escalated to the present for cost inflation), and that a price of slightly above this will at least maintain non-OPEC production while still permitting enough demand growth to result in a gradual increase in the call on OPEC oil. At the time of writing, it seems that the industry has been successful, at least for a few years, in bringing its costs below this level. This has been due to technological advances and, more recently, to the political liberalisation of low cost areas.

THE ECONOMICS AND POLITICS OF OPEC

OPEC in context

It will have become apparent from previous sections that OPEC behaviour can only be understood in the context of the market in which it operates, i.e. the demand for oil and the economics of competitive supply. As the history of its often acrimonious meetings has amply demonstrated, however, it is hardly a homogeneous unit, despite the remarkable cohesion which has permitted it to continue to function during the years when two of its members, Iraq and Iran, were at war with one another, and despite the more recent invasion of Kuwait by Iraq.

Politics following economics

As is so common, political behaviour follows economic realities, and the positions taken over time by the shifting factions within OPEC, though often appearing to be inspired by largely political differences, can in fact be traced to real or perceived economic forces. Although formed in 1960, in its early years the organisation made no attempt to control prices. Instead, its members concentrated on increasing the share which they received of the profit attributable to oil production.

Early years

By the time of the Yom Kippur War and Arab oil embargo of 1973 and more radical approaches to nationalisation and oil pricing, the real oil price declines of the previous 20 years had encouraged rapid demand growth and stagnation in supply capacity. The oil market was tight, and it was easy for OPEC to defend the prices which followed the first oil crisis of 1973/4. The oil market was again very stable during the remaining 1970s, with little division within OPEC.

The crisis of 1979–1981 was rather different. Even at the time there were sharp differences between the 'hawks' – notably Iran, Libya and Algeria – who were calling for OPEC official prices to reflect fully the peaks hit by spot prices at the height of the panic, and the 'doves', notably Saudi Arabia and the other Gulf Arab states, who, with Venezuela acting as a mediator, argued for price moderation.

Internal divisions

Although the composition of the debate has changed slightly over time, the positions taken by OPEC members are driven by the same factors. Members with long oil reserve lives, low populations and stable finances have preferred price moderation. Members with short reserve lives, large populations and unstable finances have preferred to maximise short term prices.

Effect of changing opinions

Perceptions of what is sustainable did change, however. It was with extreme reluctance, in 1981, that Saudi Arabia agreed to as high a price as £34/bbl for Arab Light. Had it known that in its attempt to defend this price it was to reduce its output from 10 mmbbl/d to $2^1/_2$ mmbbl/d within 5 years, it would certainly never have accepted it. Since 1986, the OPEC member with by far the greatest spare capacity has been Saudi Arabia, and it has been this factor, in addition to the size of its output, which has re-established Saudi domination of OPEC pricing policy.

Pricing system

The system of OPEC pricing which has been described depends on the assumption that, firstly, Saudi Arabia and the other moderate Arab states continue to believe that pricing at or just above the long run marginal cost of non-OPEC supply remains in their long term interests, and secondly, that they continue to maintain a manageable but adequate buffer of surplus capacity between themselves.

Disequilibrium

Excess demand

With all producers, inside and outside OPEC, producing at full capacity, it is in effect impossible to bring significant numbers of additional barrels onto the oil market if it gets into short supply. The only market adjustment mechanism would then be via the effect of higher prices choking off demand. However, even on the demand side of the supply and demand balances, sensitivity to price is fairly low in the short term. This creates the theoretical possibility of very high prices in the short term, even though these would clearly be eroded again in the slightly longer term. It is for this reason that the signatories to the International Energy Agency (essentially the OECD group of countries) agreements carry higher than required inventories, and, notably, why the US government has crude oil stockpiled in the Strategic Petroleum Reserve (huge stocks bought by the government and pumped into underground storage). While these storage arrangements are adequate to cope with the effect of a short term supply disruption and could add several million barrels per day to immediately available supply, they clearly have only a limited life. If there is to be a long term trend of stable, rather than rising, oil prices, adequate production capacity will need to be maintained.

Excess supply

Surplus capacity within OPEC creates the constant possibility that its owners will use it, driving prices down to the level at which they create additional demand or cause higher costs suppliers to stop producing because they are not covering their production costs. This would require extremely low, and unstable, prices. Thus, a stable OPEC policy actually has two elements: the management of supply such as to keep prices stable at the required level, and the maintenance of capacity at a level which is more than adequate to meet peak demand, but not so substantial as to make wars for market share an inevitability.

For oil as for other commodities, there has historically been a high correlation between capacity utilisation rates and the direction of real prices. Since the non-OPEC countries always produce to capacity under the umbrella of OPEC pricing, it is in fact the extent of unutilised OPEC capacity which is the key statistic.

OPEC capacity utilisation

The 'old rule of thumb'

The two oil price shocks of 1973–1974 and 1979–1981 occurred at times when OPEC was operating at very high utilisation rates; over 90% in both cases. During the early 1980s, when OPEC was trying to support impossibly high prices, its utilisation rate fell to below 70%. The rule of thumb was that there was a comfort level of about 80% which was compatible with stable prices; higher, and there was the increasing risk of a crisis, lower, and there would almost inevitably be some downward drift.

Invasion of Kuwait

The above analysis was always unconvincingly mechanical, and it was invalidated by oil market behaviour during and after the Gulf crisis of 1990–1991. When Iraq invaded Kuwait in the summer of 1990, it simultaneously shut in Kuwait's production of 1.5 mmbbl/d and its own production of 3 mmbbl/d, at least as far as international markets were concerned, since it was immediately embargoed from exporting its own production.

Although oil prices rocketed ahead of Operation Desert Storm, they collapsed back to pre-invasion levels during the war itself, once it became obvious that Saudi exports would not be interrupted. Throughout the subsequent year, until Kuwait's production recovered to reasonably high levels in 1992, the oil market operated with OPEC at over 95% capacity utilisation, and yet most of the risks to the oil price were on the downside; in the spring of 1992 it was even necessary for OPEC to undersupply the market briefly, to restore required prices. Its failure to repeat this in 1993 led to pronounced oil price weakness.

Reasons for change

There seem to have been two reasons for the change in market behaviour. The first was that the oil market of the post-war period was overhung by the knowledge that room would have to be made for Kuwait and, eventually, Iraq, so there was always the risk that a sudden political event would spin the market from being tight to confronting an OPEC war for market share.

The second, equally important, point was that oil companies and traders had learned from experience. In none of the three supply panics (1973, 1979 and 1990) had there actually been a serious shortage. After the third crisis there had not even been any attempt by OPEC states to sustain prices at their peak level, or to prevent them from falling to pre-crisis levels. Consequently, there was a much lower incentive for oil traders and companies to hold stocks. In the oil market, as in other markets, participants do not always respond in the same way to the same stimulus, and extrapolation from past experience is often inappropriate.

'Supply crunch' fears

Whatever the change in the market's perception of what is an adequate level of surplus capacity, it remains the case that the relationship between expected demand for OPEC oil at current real prices and OPEC plans for capacity expansion should bear some relationship with one another if oil prices are to remain stable.

Ahead of the recession of 1990–1992, it was becoming fashionable to argue that there might be an OPEC 'supply crunch' in the mid-1990s. The point was that there had been no need, since the 1970s, for OPEC countries to add to their capacity. The ability and cost of OPEC members doing so was uncertain, as was the issue of how this would be funded.

The Gulf War changed the nature of this debate somewhat. Saudi

Arabia added a million barrels per day of capacity fairly effortlessly (albeit by de-mothballing a field rather than by constructing significant new facilities), and confidence in the ability of the Gulf countries to increase their production was enhanced. Moreover, the extremely attractive economics of adding capacity in the Middle East, as against non-OPEC areas, was underlined by a new willingness on the part of Iraq, in negotiations primarily with the French oil companies, to offer contractual arrangements which effectively would permit equity participation for the corporate partners (once, of course, the United Nations embargo on Iraqi exports was lifted).

Funding requirements

If the physical ability and cost of adding to Middle Eastern production capacity had become less of an issue by the post-war period, the financial ability of the countries concerned to fund such expansion was in greater doubt. OPEC members' finances ranged from the barely manageable to the chaotic. The huge petrodollar deposits of the 1970s had been eroded by the end of the 1980s. After the Gulf War, OPEC Middle Eastern states, with the single exception of the United Arab Emirates, were debtor nations.

Despite the concerns, however, it seems extremely probable that OPEC states will make most of the planned additions to capacity which will be required to keep up with demand through the remainder of the 1990s. In some cases, notably Iraq, this will ultimately require the admission of foreign capital. Most OPEC members will probably be able to borrow the required funds. In most cases the returns on the investment will be high, and the extent to which they are politically motivated should not be neglected.

Reconstruction

Saudi Arabian production peaked at 10 mmbbl/d in 1980, and resumption of that level of capacity was as much a matter of national prestige as OPEC politics. For Kuwait and Iraq, restoration of pre-war capacity has been a matter of priority, irrespective of cost. And even Iran refers back to its pre-revolutionary position as the second largest player in OPEC (and political rival to Saudi Arabia's leadership in the Gulf).

An example forecast

Table 2.1 illustrates estimates made in 1992 by Mr Ian Seymour, Senior Editor of the *Middle East Economic Survey*, of realistic OPEC capacities by country in 1995 and 2000. For illustrative purposes, these are related to what were then consensus figures for oil demand and projected non-OPEC consumption. The message was mixed. Utilisation rates would remain high by historical standards, but a large and increasing sur–plus capacity would emerge during the first half of the 1990s, which would require managing if it was not to depress prices. It was against this background that the expectation of flat real prices, at best, was formed.

Table 2.1 *OPEC supply/demand/capacity projections, mmbbl/d*

	1991	*1995*	*2000*
World oil demand	66.4	70.0	73.8
Non-OPEC supply	41.4	40.5	39.5
Call on OPEC oil (including NGL)	25.0	29.5	34.3
Call on OPEC crude	23.0	27.4	32.0
OPEC crude capacity	24.5*	33.3**	37.2
OPEC excess capacity	1.5	5.9	5.2
% Capacity utilisation	93.9	82.3	86.0
% Gulf share of OPEC capacity	64.2	70.0	73.4

* *Iraq (export constrained) at 0.42 mmbbl/d and Kuwait at 0.4 mmbbl/d.*
** *Iraqi and Kuwaiti capacity restored to pre-war levels of 3.2 mmbbl/d and 2.4 mmbl/d respectively.*

OPEC crude production capacity projections, mmbbl/d

	1991	*1995*	*2000*
Algeria	800	800	600
Ecuador	300	300	250
Gabon	300	300	250
Indonesia	1 450	1 400	1 300
Iran	3 400	4 500	4 500
Iraq	420	3 200	4 500
Kuwait	400	2 400	3 000
Libya	1 600	1 800	1 800
Nigeria	1 900	2 400	2 500
Qatar	400	400	300
Saudi Arabia	8 700	10 000	12 000
UAE	2 400	2 800	3 000
Venezuela	2 400	3 000	3 200
Total	24 470	33 300	37 200

Source: Middle East Economic Survey, *1992.*

Conclusions

Circumstances, and therefore appropriate forecasting methodologies, change over time. So long as OPEC policy remains as it has been since 1986, and ignoring short term disruptions, we suggest that crude oil price forecasts should place greater weight on evidence as to the full cost of non-OPEC supply, and should assume that this represents the bottom of a range within which Saudi Arabia will attempt to manage prices. Its ability to do so will probably depend on this price being compatible with the continued expansion of OPEC capacity.

Intuitively, given the tendency for costs to be lower in the Middle East than in the main non-OPEC areas, this should not represent an unworkable constraint. On the other hand, downward pressures on oil prices, perhaps to below the non-OPEC long run marginal cost, are likely as large additions to capacity are made. The return to the oil market of Kuwait and Iraq represents a case in point. Over the longer term, however, this lower price must, by definition, result in demand swinging back to OPEC, and absorbing the surplus capacity. The greatest risk is that failure to support a high enough price results not only in underinvestment outside OPEC, but also deferral of additions to Middle Eastern capacity. This combination would, eventually, create a supply shortfall.

THE SHORT TERM

Introduction

This book is about the evaluation of oil companies and assets. It is concerned with the history of and ways of looking at external variables – such as the dollar crude oil price – only to the extent that these are inputs to any evaluative process for oil company assets. It follows that the short term volatility of the oil market is largely irrelevant to this exercise, and that it is the long term trends on which the analyst should concentrate.

It is, however, necessary to be able to differentiate between ephemeral and underlying factors, and overall understanding of the oil market requires some knowledge of its day-to-day operation. Moreover, all interested parties (investors, oil company issuers of or acquirers of shares, and even oil company traders in assets) will try to finesse the timing of their activities in such a way as to obtain the best terms for themselves, even if they also make their strategic decisions on the basis of expected long term trends. We shall therefore devote a little space to the issues specifically surrounding short term oil market volatility.

Quarterly periods

It is conventional, when projecting short term supply and demand balances, to do so using quarterly periods. The reason for this is that there is a sharp seasonal fluctuation in oil demand, which peaks in the Northern hemisphere winter and is at its lowest in the spring. Within this overall pattern, there is also some shift in the slate of oil products required by season. The winter quarters reflect a demand for gas oil for home heating purposes, and the summer tends to show an increase in gasoline consumption, especially in the US.

Inventory

Time horizon
Whereas in the long term it is reasonable to ignore fluctuations in oil industry inventory, concentrating instead on costs and capacity, in the short term they generally have the greatest impact on oil prices.

Market reactions
Oil companies and traders use models which project demand and non-OPEC supply. They then take a view on what OPEC sales are likely to be (in many cases this requires subjective assessment as to the level of OPEC cohesion and what individual producers' priorities are), and derive a projected movement in oil inventory. Depending on how this relates to normal inventory movements and the adequacy or otherwise of current stock levels, the individual trader will then take a view on the likely next move in oil prices.

OPEC market management
OPEC, in its deliberations, takes the above process in reverse; that is, it derives a forecast demand for oil from its own sales and from inventory,

and then sets its production quota at the level which implies a movement in inventory which is likely to cause oil prices to move in the direction which it desires.

Inventory is usually measured by the oil industry in terms of future supply days covered, and it is the industry inventory, rather than the government held strategic stockpiles, which drives the price. Table 2.2 illustrates the supply and demand balances produced by the OPEC secretariat to advise ministers at the October 1992 meeting, at which it was decided to fix the overall OPEC winter sales quota at a level which implied probable sales of about 24.7 mmbbl/d, which was then regarded as a level compatible with stable oil markets. As intended, spot oil prices were left virtually unchanged by the outcome of the meeting. As is so often the case, subsequent events upset both the supply and the price calculations, but they were reasonable at the time when they were taken.

Uncertainty

Unfortunately, the situation is not quite as simple as the above would suggest. Oil industry statistics are inaccurate to an extent which exceeds the movement in inventory which drives oil prices. Thus, far from supply and demand balances being a reliable guide to future oil prices, it is all too often that oil prices provide worrying evidence that oil companies and OPEC have over- or under-estimated demand.

Even the historical data on which the International Energy Agency bases its monthly *Oil Market Report*, the most reliable published source of short term forecasts, is often heavily revised. Apart from the accuracy or otherwise of historical data, there are two other fundamental problems with the simplified model of the oil market represented by quarterly supply and demand balances:

1 Demand for crude oil is derived from demand for oil products, and it is often the supply and demand balances for individual products in individual markets which have a strongly determining influence on oil prices.

2 There are substantial time lags built into the oil market system, and these have the effect of exacerbating the oscillations which lack of accurate data would in any case make inevitable.

Table 2.2 *World crude oil supply/demand balance, mmbbl/d*

	1988	*1989*	*1990*	*1Q91*	*2Q91*	*3Q91*	*4Q91*	*1991*	*1Q92*	*2Q92*	*3Q92*	*4Q92*	*1992*
World crude oil requirements[1]													
OECD[2]	36.38	36.73	36.60	37.10	35.49	35.68	37.66	36.48	37.68	35.73	36.61	38.20	37.06
Former CPEs[3]	12.47	12.51	12.13	12.43	12.00	11.59	11.61	11.91	11.50	11.35	11.25	11.20	11.33
Others[4]	14.01	14.71	15.38	15.78	15.76	15.96	16.08	15.90	16.41	16.49	16.55	16.65	16.52
Total	62.86	63.94	64.11	65.31	63.25	63.23	65.35	64.29	65.59	63.57	64.41	66.05	64.91
Non-OPEC oil supplies and world NGL crude oil production[5]													
OECD	14.36	13.66	13.61	14.12	13.51	13.73	14.14	13.87	14.22	13.81	13.81	14.09	13.98
Developing countries	8.58	8.80	9.20	9.47	9.52	9.41	9.49	9.47	9.56	9.63	9.71	9.47	9.66
Former CPEs[6]	15.50	15.20	14.44	13.83	13.53	13.18	13.02	13.39	12.70	12.33	12.71	11.87	12.27
Non-OPEC NGL[7]	2.71	2.71	2.74	2.90	2.90	2.90	2.90	2.90	3.10	3.10	3.10	3.10	3.10
OPEC NGL	1.70	1.88	1.87	1.85	1.85	1.90	1.92	1.88	1.93	1.94	1.97	1.99	1.96
Total	42.85	42.25	41.86	42.17	41.31	41.12	41.47	41.51	41.52	40.81	40.76	40.79	40.97
Required production from OPEC and stocks	20.01	21.69	22.25	23.14	21.94	22.11	23.88	22.78	24.07	22.76	23.65	25.26	23.94
OPEC crude oil production[8]	19.54	21.66	22.97	22.96	22.56	23.54	24.04	23.27	24.08	23.66	24.07**		
Balance	−0.47	−0.03	0.72	−0.18	0.62	1.43	0.16	0.49	0.01	0.90	0.42		

OECD opening stocks, mmbbl										
Commercial onland	2685	2606	2595	2688	2625	2659	2764		2663	2552	2555[e]
SPR	753	795	832	845	828	828	834		843	847	848[e]
Total	3438	3401	3427	3533	3453	3488	3599		3506	3400	3403[e]
Days of forward consumption in OECD											
Commercial onland stocks	72*	69*	69*	72	74	75	73		71	71	70
SPR	20*	21*	22*	23	23	23	22		22	24	23
Total	93*	90*	92*	95	97	98	96		93	95	93

Note: Totals may not add up due to independent rounding.

1. *Defined as: Inland deliveries of petroleum products + bunkers + direct use + refinery own use − refinery processing gains − backflows.*
2. *OECD includes the former East Germany.*
3. *Apparent consumption (production − net oil exports) for the former USSR (still including the Baltic Republics), China, Albania, Bulgaria, Czechoslovakia, Hungary, Poland and Romania.*
4. *Includes the demand in Yugoslavia, Cuba, Mongolia, Vietnam, Laos, Cambodia and North Korea as well as OPEC countries and all other developing countries. Excludes non-conventional oils in developing countries.*
5. *Includes non-conventional oil produced only in OECD. Excludes such oils in non-OECD countries.*
6. *Includes NGL.*
7. *Excludes former CPEs' NGL.*
8. *Selected secondary sources.*

e *Estimated.*
* *Applies for first quarter of the year only.*
** *July average.*
Source: OPEC Bulletin, *September 1992.*

Individual products

Customers of the oil companies do not burn crude oil (except in Japan, where it is used as an alternative to fuel oil and burned directly). Spot product prices reflect the adequacy or otherwise of supplies of individual products in local markets, because while crude oil is cheap to transport, product is not, and in any case it cannot be transported instantaneously.

Thus, in the spring of 1988 and 1989, it was the potential for short supply in the US gasoline market which drove the spring rallies in oil prices. In the winter of 1990–1991, it was the lack of middle distillates (kerosene and gas oil) in the Far Eastern markets, following the closure by the Iraqis of the Kuwait export refineries, and US military demand, which were responsible for the heights reached by oil prices. In the winter of 1991–1992, it was the high level of European and American distillate inventory, combined with a mild winter, which caused oil prices to break downwards. These price disruptions were considerably greater than analysis of the overall global oil balances would seem to have justified.

Time lags

Time lags arise for two reasons: firstly, because of the time involved between the nomination of production volumes – especially in the Middle East – and their arrival in Western markets, and secondly, because of the time involved between the accumulation of or depletion of inventory and its being recorded in published figures so that the industry can react to it.

The only geographical market in which industry stock data is available regularly is the US, in which it is collected weekly and the previous week's closing balances are generally published by the America Petroleum Institute after markets close on the subsequent Tuesday. In Japan and most European markets, data is only collected monthly. In other markets, data is often collected infrequently and erratically.

More important than the problem of data collection, however, is the lag associated with so-called 'long haul crude'. Much non-OPEC supply is produced fairly close to its markets. North Sea oil, for example, is geographically close to the refining complexes of north-west Europe, and US oil, other than that produced in Alaska, can usually be piped directly to the plant at which it is refined. This does not apply to Middle Eastern oil. In general, sales from the Middle Eastern OPEC producers are made under volumes nominated in the month ahead of the actual delivery, which is made FOB (free on board) at the loading terminal. It then takes several weeks to transport the crude oil by tanker to its target market. During this time, the risk of price fluctuation is borne by the oil company, or trader, that owns the oil while it is in transit.

Paradoxically, however, it is the long haul crude from the Middle East which, as we have seen, fulfils the role of swing supply. The base load, produced to capacity throughout, is the non-OPEC oil which is geographically the closest to its target market. Thus, not only does the oil industry deplete the more expensive oil first leaving cheaper Middle Eastern oil in the ground; it also attempts to fine tune the supply of oil onto the market by regulating the oil which takes the longest possible time to arrive, and continues throughout to produce to capacity that for which the effective supply could be regulated quickly.

OPEC countries, notably Saudi Arabia and Iran, have in the past accumulated floating storage in Europe and the Caribbean so as to be able to sell it as effectively short haul crude oil, but they have only a limited incentive to undertake the cost associated with this exercise in reducing volatility, and this activity has thus been intermittent.

PRICE

Price setting

Despite the role of long haul crude as the swing supplier, its price is generally not transparent, and is set under contractual arrangements with the oil companies which link it to the spot prices of Western crude oils. This is an additional paradox of the industry, that while overall prices are regulated by the supply of swing Middle Eastern supplies, the price of those supplies is fixed with reference to the transparent spot price of the short haul crudes, which are always produced to capacity, but whose prices react to the surplus or deficit in the overall crude oil market.

In effect, the price of Saudi Arab Light crude is set in relation to US West Texas Intermediate (WTI) and UK Brent, but the price of WTI and Brent is set by the volume of Arab Light which is supplied. This is because the markets on which some Western crude oils trade are extremely transparent and have a liquidity which is disproportionate to the scale with which the crude oils concerned are produced.

The markets

The need for markets

Not all markets actually involve the purchase and sale of physical barrels at all. Structurally, the oil industry is far more complex than a description of the bigger integrated oil companies would suggest. Companies are not balanced in such a way as to produce, refine and market exactly the same volumes at each stage in the chain, so there is an active trade between them. For large customers, oil costs may be a large proportion of their operating costs, e.g. airline companies, and they therefore have a requirement to fix their costs forwards as far as is possible. Finally, the level of tax which applies to the different stages in the oil industry chain, notably the upstream production business as against the downstream refining and marketing business, varies enormously (in general upstream taxation is much heavier).

As with other industrial conglomerates, oil companies have also been under increasing pressure from their shareholders to perform, and this has resulted in a shift in priorities from long term strategic considerations to immediate profitability. Internal transactions have thus become 'arm's length', not just because of the pressures applied by revenue authorities.

Major markets

In addition to active spot markets for crude oil and oil products which have emerged at most of the world's refining centres, there are also two main markets which permit traders to hedge positions forward. These are the North Sea 15-day Brent market, and the New York Mercantile Exchange (NYMEX).

Brent market

The Brent market is a hybrid, in that although most of the transactions undertaken in it are 'paper transactions' (there is no necessary intention to take physical delivery), the contract is in cargo loads of crude oil delivered at Sullom Voe terminal from the Brent system. Vessels have to be nominated for loading a fortnight prior to delivery, so the contract to purchase or sell abruptly switches from a 'paper' to a 'wet' barrel, once the nominations have been made.

Prior to this point, however, Brent cargoes are often traded several times (in so-called 'paper chains'), and are customarily used as hedging instruments by parties which have no intention of taking actual delivery.

The market is regulated by the major oil companies involved, which have drawn up standard contracts between traders, but it is not a futures market in the conventional sense; it is a forward market.

The NYMEX

In addition to Brent, the other main Western market is the NYMEX, which trades in conventional futures contracts covering small parcels of both crude oil and products. Although the NYMEX contracts do permit physical delivery, the vast majority of contracts are closed for cash prior to their expiry. In addition, the NYMEX is a regulated futures market which underwrites the transactions which are taken on it.

Liquidity

Despite their differences, however, the two markets are alike in establishing transparent and liquid markets in oil for forward delivery. The scale of the markets should not be underestimated. At the height of the Gulf crisis of 1990/91, turnover on the NYMEX was over 100 mmbbl/d, against total world consumption of about 66 mmbbl/d. Although oil industry analysts can by and large ignore the mechanisms of the markets on which oil prices are set, company policies with regard to hedging are significant to their profitability, and third party trading does distort the booked refining and marketing profits made by integrated companies.

REFINING MARGINS

Introduction

Crude and products

If the exploration and production businesses of oil companies are driven by crude oil prices, so the refining and marketing businesses are driven by the margin between the value of a barrel of crude oil and the value of the products into which it is transformed.

Refining and marketing

The downstream oil margin can be split into two elements. The first is the spread between the spot crude oil price and the wholesale value of the slate of products into which it is refined, minus the cost of operating the refinery and of transporting the crude oil to the refinery (known as the 'netback value' of the crude oil). The second element is the further margin between the wholesale value of the individual petroleum products and the net of tax inland price at which the oil company markets them through its retail system.

Refining margin volatility
A more detailed description of the technical processes and the commercial structure of the downstream end of integrated oil companies is provided in Chapter 5, but the immediately important conclusion at this point is that whereas there are underlying differences between the downstream profitability of different oil companies, the oscillations in their downstream profits are to a large part explained by fluctuations in the refining margin in their regional area of operation.

Regional markets
The regional distinction is important here, while it is not in the crude oil market; the reason is that while it is extremely cheap to ship crude oil, it is much more expensive to transport finished petroleum products. Many of these, mainly the transport fuels, are highly inflammable and have to be kept free from contamination if they are to be delivered to specification. Hence the distinction between shipping 'white products' and 'dirty products' in quotations for tanker chartering. The white products include the gasolines, kerosenes and gas oils, and the dirty products are the fuel oils and crude oil. Parcels of white products are typically much smaller than of dirty products, which explains why the oil majors used to build refining capacity close to consumer markets, rather than close to the producing fields.

Refinery locations
It remains the case that the bulk of refining capacity is close to target markets, but this began to change in the 1970s, as OPEC members began to build export refineries and petrochemical plants. To some extent, there was a political rather than merely commercial justification for the siting of these units. OPEC members have, on the whole, less developed economies, and domestic downstream integration is seen as a stage in their economic transformation. (This should not be allowed to obscure the fact that downstream overseas investment by OPEC members such as Venezuela, Kuwait and Saudi Arabia has been significant. In addition to their economic development, OPEC members have also integrated downstream for commercial reasons.)

A second factor, which has emerged more recently, is the pressure of environmental legislation in the OECD countries, which has made the construction of 'greenfield' downstream sites effectively impossible. Capacity continues to creep, through the de-bottlenecking of existing plant, but it is almost inevitable that future oil product demand growth, notably in the US, will be met by increased volumes of imported oil products. Even so, it seems likely that the regional markets of the Far East, Europe and the US will remain relatively independent, though of course there are connections between them.

Regional histories

An illustration of the relative independence of the different geographical regions was offered by the trend in refining margins through the Gulf crisis. Iraq's invasion of Kuwait resulted in the shut-down of Kuwait's refining plant, which in turn had been an important source of supply to Far Eastern markets. Oil product prices in the Far East rocketed, widening refining margins there to record levels. Middle Eastern margins rose considerably as all the remaining plants in the area operated at full capacity to attempt to meet local and Far Eastern needs. For a brief period, rather than Middle Eastern product flowing into the Mediterranean area to help meet European needs, the opposite occurred – product flowed south through the Suez Canal, which in turn pulled up European margins. Meanwhile, partly under pressure from the US government not to increase gasoline prices at a time of war, and partly because of the recession, US margins remained extremely depressed. It is unlikely that in time they would have remained unaffected by events elsewhere, but with cessation of military operations in the Gulf, and further recession, refining margins in Europe came off the boil.

Margins interpreted

Driving factors

Despite their geographical differences, the factors which drive refining margins are essentially the same internationally. The situation is complicated by the fact that different crude oils and different refinery configurations permit the production of different product slates; so, in reality, the structure of the margins of different refiners in a single location will differ, and a given price movement will be of more benefit to one than to another. This issue is covered in some detail in Chapter 5, but for all of these complications it remains the case that the more significant shifts in refining margin tend to affect all refiners to a greater or lesser extent, and that they can mostly be ascribed to one or another of two causes.

Figure 2.6 illustrates the trend in European refining margins using half year averages over the period 1981 to the first half of 1993. The margin is expressed in French francs per tonne (the source for the data was Total), and has been adjusted into 1992 currency. In other words, the margin shown has been adjusted for the effect of inflation and illustrates the real trend in European refiners' margins over a period of more than a decade.

Crude price volatility

Crude price volatility is notable, and explains the violent fluctuation in the profits of refining and marketing companies, even prior to the destabilising effect of inventory value increases and decreases. It is also notable that the most violent oscillations in refining margins have been associated with periods, such as the summer of 1981, the spring of 1986, and during the Gulf War, when crude oil prices were moving violently.

In 1986, the collapse in the crude oil price preceded a fall in product wholesale prices (in turn these took time to be passed on in inland prices) and the result was a huge increase in margins. Hence the phenomenon of 1986, in which oil companies' refining and marketing profits were unusually large prior to stock losses, though very depressed

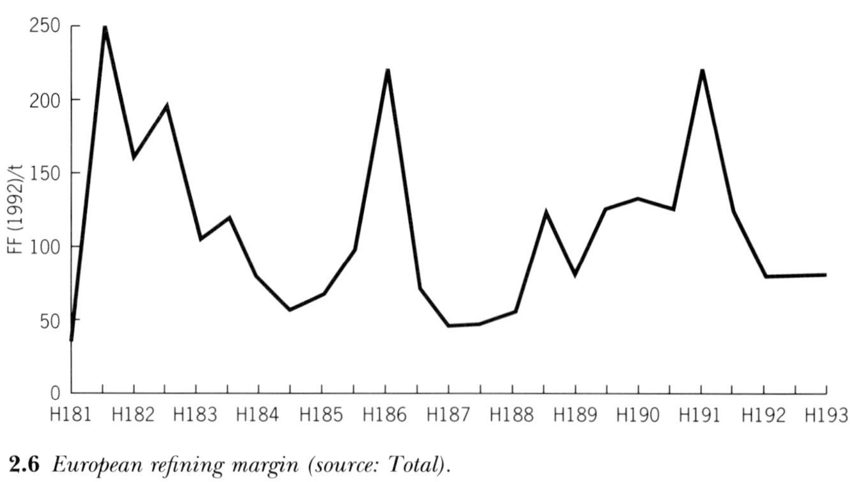

2.6 *European refining margin (source: Total).*

when the effect of inventory value reduction was taken into account. The significance of stock effects to oil companies, and how they should be interpreted, is covered in Chapter 3, but it is clear that the margin available to refiners at times of oil price turbulence is often unrepresentative of underlying industry conditions.

During the Gulf crisis, the initial effect on oil prices was for the more volatile and speculative crude oil price to run upwards, driving down refining margins. Subsequently, as it became apparent that there was no problem with crude oil availability, the refining squeeze described above came into effect; the crude oil price fell but product prices rose. Refining margins soared.

Refinery utilisation

Even ignoring the two periods of turbulence in 1986 and 1990, it is evident that there was also an underlying trend to the margin over the period. After the upheavals of the 1979–81 period, refining margins fell steadily during the first half of the 1980s. From their level in 1987, after the 1986 price crash, European refining margins strengthened steadily through to the first half of 1990. When the effects of the Gulf crisis ended in the first half of 1991, they fell again, and only stabilised late in 1991 at relatively low levels which were subsequently maintained.

Clearly, there is a floor to refining margins, and it is the level at which plant is shut in with operating margins leaving no surplus over operating costs. There is also, in the longer term, a ceiling, which is the level at which large amounts of new capacity are commissioned, because the cost is justified by current or expected operating profits. The last time that large amounts of new plant were installed in the refining industry was during the 1970s when, as we have seen, there was a universal expectation that demand for oil products would grow, whatever the oil price. In the event, demand fell during the early 1980s, and much refining plant was shut down.

The period of the late 1980s, through to the recession of 1990–92, was by contrast characterised by increasing throughputs and stable capacity. Moreover, margins, though widening, were still not adequate to support new projects. Only in the Far East was the expectation of further demand growth, and the reality of wider margins than in Europe and the USA, adequate to induce companies to plan new greenfield refineries. Figure 2.7 illustrates the history of OECD refining plant utilisation rates over the 11 years, 1981 to 1991.

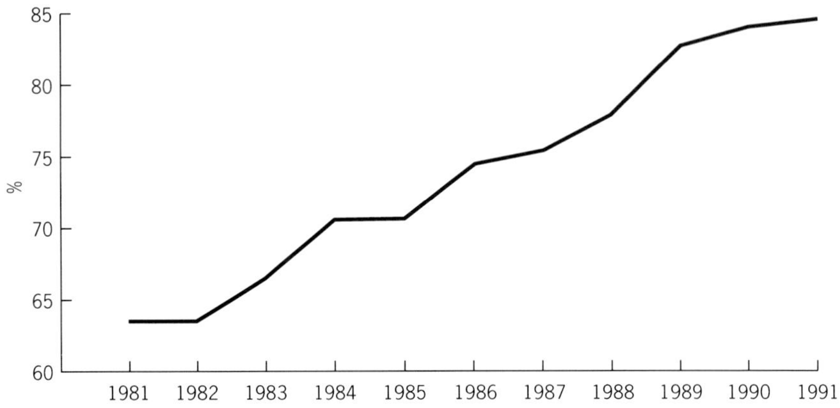

2.7 *OECD refinery utilisation rate (source:* Statistical Review of World Energy, *British Petroleum, 1992).*

Environmental legislation

A new influence

Over and above the factors of refinery utilisation rates and short term disequilibria (whether due to oil crises or plant closures) there is a new factor which is increasingly imposing itself on the refining industry, and this is environmental legislation. We have already referred to this in commenting on the location of refineries, but it is also having an increasing impact on existing plant.

European restrictions

Environmental legislation has been aimed at reducing the damage caused by emissions from the fuel burning process and at reducing the direct environmental impact of refineries. In the European Community, it has primarily been restrictive, limiting, for example, the sulphur content of diesel fuels and heating oils. Differential excise tax rates have also been used to encourage switching to unleaded gasoline, and a requirement for catalytic cleaning of exhaust emissions is being phased in. It is probable that the benzene content of gasoline will, in time, also be restricted.

US restrictions

The trends in Europe follow, and are a good deal less draconian than the amendments to the Clean Air Act and the various state level provisions which have been introduced in the US. Certain US cities, notably Los

Angeles, have for some years suffered from a serious smog problem which has been exacerbated by gasoline emissions.

Despite the fact that the greater part of air pollution is attributable to older vehicles, and that the marginal benefit from greater restrictions on emissions from new vehicles is probably not commensurate with the cost of imposing them, the US has introduced legislation which is having the effect of increasing dramatically the costs for refiners (both in producing to specification and in restricting the emissions from their plant) and also of imposing tight requirements on motorcar manufacturers as they phase in new models. Furthermore, the legislation is in many cases prescriptive, rather than relying on limitation. In the case of gasoline in particular, the principle of so-called 'reformulated gasoline' is such that the legislation virtually prescribes the chemical composition of the fuel. There have also, unfortunately, been developments which probably reflect the pork-barrel nature of the US electoral system rather more than the dictates of welfare economics, such as the apparent determination to subsidise ethanol based fuels.

Traditional strategy

It is probably too early to estimate fully the effect of environmental legislation on refiners, quite apart from the indirect impact of higher prices on lower customer demand. One point is being demonstrated clearly by the US experience, however. It is that the requirement to undertake substantial capital expenditure requires a new strategy from the refiner.

Throughout the 1980s, refiners ran their plant so long as it generated positive cash flow. Whether it was profitable or not was immaterial, since the capital costs of the plant were sunk, and there was no shortage of capacity. The only investment decisions required were over how much upgrading plant to construct, and rapidly widening product price differential justified most of these incremental decisions (see Chapter 5).

Investment decisions

Environmental legislation changes the above position dramatically. If it becomes impossible to run old plant for cash and its continued operation requires large sums (typically hundreds of millions of US dollars per refinery) to meet the new standards then it becomes crucial that margins justify the investment.

Ahead of the implementation of the provisions of the Clean Air Act, this is the decision with which US refiners were faced. It was a typical application of mathematical Game Theory, since if they all invested to upgrade their refineries, none would generate an adequate return. If few did, the returns would be colossal. In essence, stringent refining legislation has had the effect of transforming refining from a mature to an immature industry, with respect to investment decisions. This process can be expected to continue, and to apply in Europe to an increasing extent.

Conclusions

Utilisation rates

From these considerations, it is evident that there are certain factors which an analyst forecasting refining margins should consider, whether the object is oil equity evaluation or industrial project evaluation. Other than to the extent of short term fluctuations, which can be quite dramatic, the main factor which has driven and which is likely to continue to drive refining margins is the rate of utilisation of basic crude oil distillation capacity. In an environment in which margins do not cover the cost of new building, this is mainly a function of trends in oil product demand.

Upgrading returns

In reality, since refiners do more than simply distil crude oil, there is also an element of return to other types of plant, notably upgrading equipment – the return to which is a function of the spreads between heavy and light oil product prices, rather than those between crude oil and products. Much of the improved margins of the late 1980s was attributable to widening of upgrading margins, rather than distillation margins (see Chapter 5).

Geographical diversity

Trends in refining margins are global, but can diverge quite considerably by region, the limiting factor being the cost of transporting products between one refining centre and another. Broadly, the main refining centres are the Far East, Europe and the US; the Middle East exports product to both the Far East and to Europe, thus connecting price between those markets. The US is effectively split between two markets, east and west of the Rocky Mountains; the west coast of the US is more closely integrated with the markets of the Far East than it is with those of the east coast and the Midwest, again because of cost of transport.

The future

Refining margins fluctuate between break even cash costs and the level which justifies new building. Outside the Far East, it has been the former, rather than the latter, which has been a determining factor, though this was beginning to change rapidly in the late 1980s. Ultimately, it seems inevitable that the refining margins of the European and US markets will have to rise from their levels of the 1980s, unless demand is to be almost static, since no new plant can be justified at present margins, and volume creep, through de-bottlenecking, is slow.

An offsetting factor, however, would be significant additional volumes of Middle East export plant, especially if this were to be built for political reasons. Although environmental legislation has the effect of reducing demand, the choices which it forces on refiners may have the effect of accelerating the trend to higher margins. If not, then returns on capital will become even less adequate.

In the Far East, the situation is different. The danger may be that demand growth is overestimated, and that excessive new capacity is introduced, but this is currently the only geographical area with inadequate plant and rising demand volumes.

NATURAL GAS MARKETS

Basic features

Transport

An important feature of the world crude oil market is that it is characterised by relatively low transport costs, as is the fact that the use of oil as an energy source commits less capital expenditure than for other fuels. This explains why crude oil prices tend to move consistently internationally, and why the developing world has a high, and fairly price insensitive, propensity to burn oil. Neither of these considerations apply to natural gas. Although it is produced in the same way as oil, gas can only be transported by pipeline, which requires dedicated capital and commits it to one customer, or it can be liquefied for transport in liquefied natural gas (LNG) carriers, but since the terminals and the vessels for this process are expensive it is again necessary to have sold the gas under a long term contract over which to amortise the capital expenditure before the project can go ahead.

This basic economic difference accounts for the extraordinary differences between gas markets internationally and the markets for heating oils and fuel oils with which it mainly competes (gas is not much used as a transport fuel). In the early days of the oil and gas industry, natural gas was regarded as a useless by-product from the exploration and production of oil. Gas had been used as a fuel for decades, in the shape of so-called 'town gas', which was manufactured from coal. This was distributed by utilities which were either state-owned or closely regulated. Gradually, natural gas displaced town gas.

In the UK, for example, this occurred only in the late 1960s and the early 1970s, with the discovery of the large gas fields of the southern gas basin of the UK North Sea, and under the pressure of the first oil crisis. Where natural gas was available from large, cheaply developed local reserves, such as in the US and parts of Europe, e.g. the Netherlands and the UK, it has been able to emerge as the major fuel supply to final energy users outside the transport sector, and it has generally been priced to be competitive with oil products.

Gas marketing

Gas has not traditionally been marketed by oil companies. It has been marketed by utilities, and these have been monopolists in the businesses of both transport and supply until very recently. While the transport of gas down pipelines is clearly a natural monopoly, since it is hardly practical to run competing gas pipelines to individual customers (though in the US there is some competition between different interstate transmission lines). This is less obviously the case for supply.

There is no clear economic reason why gas cannot be sold to customers either by the producing oil companies, or by intermediaries. Indeed, it is arguable that the effect of introducing competition into the market might result in lower prices. On the other hand, a monopolist is in a position to cross-subsidise both between one customer and another (by geographical location, for example) and less obviously by the timing of the supply. Thus, it is more expensive to supply gas on a cold winter day at peak seasonal demand than it is at periods of low demand in midsummer. The peak supply will have been met by use of gas from storage and by purchases under contracts which compensate the producer for its making available capacity which is only intermittently used (see Fig. 2.8).

In the jargon of the industry, the ratio between average offtake and peak offtake is known as the load factor. Highly seasonal demand, such as for domestic heating, has a low load factor, and is more expensive to supply than high load factor customers such as industrial customers. A competitive market will reflect this fact in its pricing. A monopolist might not.

Gas pricing

The gas industry responds to the fact that it has a steep load duration curve (as illustrated in Fig. 2.8) not only by contracting capacity which will only be utilised at times of peak demand, but also by selling gas under both firm and interruptible contracts. Thus, some large customers have the facility to switch between gas and oil, and will contract to take gas on the understanding that they can be cut off by the

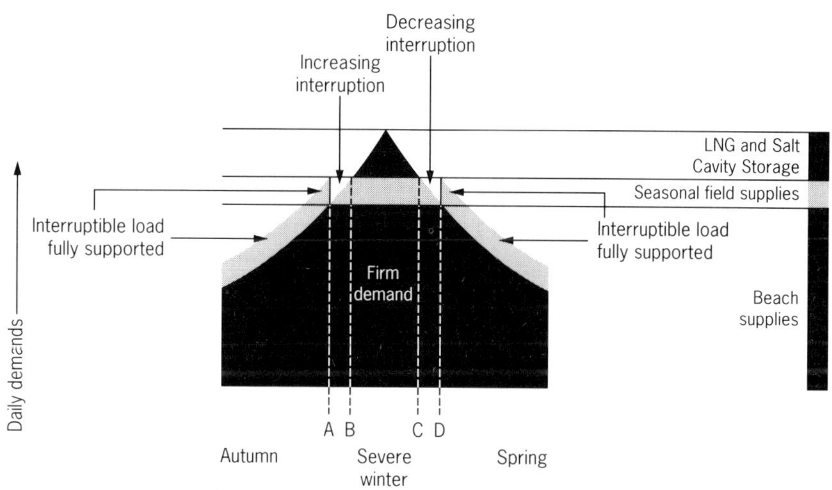

There are essentially two facets to the matching of gas supply and demand; the achievement of an 'annual' match, for the year as a whole, and of a 'peak' match, that is for daily demands in a severe winter.
The chart above illustrates a simplified peak match. In the chart daily demands are plotted in increasing order of severity up to mid-winter and decreasing from mid-winter. Firm demand in autumn is less than available beach and seasonal supply. Interruptible load can therefore be fully supported up to day A when total demand equals maximum supply. After day A decreasing amounts of interruptible load are supported as firm demand continues to rise, until at day B no interruptible load is supported. Between days B and C any further increase in firm demand is met from LNG/Salt Cavity Storage. From day C interruptible load is gradually restored until at day D all of it is being supported.

2.8 *Gas supply and demand: simplified peak match (source: British Gas).*

utility at times of peak demand. In a competitive market, the same effect occurs by the gas price rising to the point at which dual-fired capacity will choose to switch to oil. This book is about the analysis of oil companies, not gas utilities, but the wellhead price of gas to oil companies is a function of the structure of the local market into which it is competing.

Regional markets

There are three big gas markets in what used to be called the non-Communist world. The republics of the former USSR have a huge gas industry, but it has been in turmoil since the collapse of the old regime, and is of significance to Western companies only as a supplier of exported gas into Europe. The three markets are those of North America, of Europe and of the Far East. Virtually no gas flows between these markets, and they have completely different price and regulatory structures.

North America

Market structure

In the US, the gas companies divide mainly between local distribution companies and inter-state transmission companies. Gas is mainly produced in the area around the Gulf of Mexico and is also imported from Canada. Over many years, the US has dismantled the system of long term contracts which previously connected the three elements of the gas industry together, and broadly the situation is that producers are now free to market directly to large customers and to distribution companies, with the transmission companies increasingly reduced to carriers of third party gas. The local distribution companies supply gas to customers other than those which are large enough to purchase directly from producers or independent gas marketers.

Price volatility

As one would expect, the consequence for US gas prices is that they have become very much more volatile. Much of the pressure for decontrol came from the producers, who had found themselves with unsold gas in large quantities, and throughout the period of decontrol the US has tended to have a surplus of gas capacity, so prices have tended to be low, as well as volatile. This surplus is often referred to, misleadingly, as the 'gas bubble', and companies sometimes refer to the hope that the surplus will erode and permit closer price parity with oil. The reality is more complicated.

Because the demand for gas is seasonal, and the US industry balances supply and demand through price volatility, there are two possible equilibrium prices. The highest sustainable price is that at which the price of gas at the burner tip is equivalent to that of fuel oil; above this and a significant number of gas customers will switch to oil. This price nets back to a wellhead price to the oil company after deduction of pipeline charges. The lowest sustainable price to the producer is that at which producers respond by shutting in gas, in preference to producing it at the prevailing price. This may be at the marginal cost of production, or higher if they anticipate better prices shortly (see Fig. 2.9).

As a futures market in natural gas in the US has recently been created, the very short term price can also be distorted by the unwinding of positions in the futures contract. Obviously, the average price of natural gas in such a market is determined by the absolute levels of the ceiling and the floor, and by the amount of the year in which the market is characterised by gas-against-gas or gas-against-oil competition. Figure 2.9 shows a wellhead US gas price history, illustrating the point clearly, as

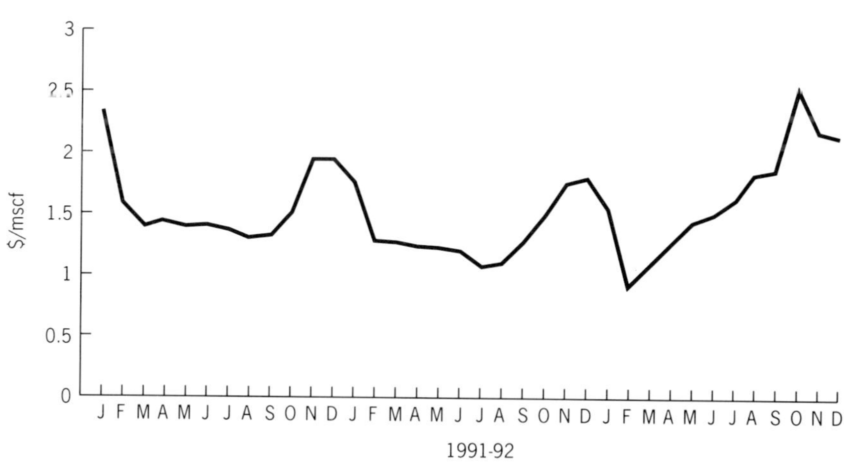

2.9 *East Texas wellhead gas price history (source:* World Gas Intelligence*).*

does the fact that during the recession the market has spent a large portion of the year at the lower end of the range, at which the producers do not recover the capital cost of capacity.

Price projections
For a given oil price outlook, projections of North American gas prices should concentrate on the relationship between peak demand and supply capacity, since it is the amount over a year at which dual-fired plant burns oil that will decide the average price for the year. It is an important feature of this market, which does not occur elsewhere, that the gas is mainly produced onshore, from reserves which can be brought into production quickly, and at low capital cost. Until the US becomes more dependent on imported LNG or offshore reserves, this factor will probably tend to keep prices averaging closer to the lower end of the range.

Europe

Structure
The European gas industry is completely different from the US, except in the UK, where there has been a determined effort to introduce competition into the energy utility industries, with rather mixed success and consequences. Most European gas markets are still dominated by a state-owned gas company, or transmission and distribution companies dealing with one another, and with the oil companies, under long term contracts.

Much of the gas that is supplied in, for example, France and Germany is imported, and the industry is in part characterised by its import arrangements. There are two large producers of surplus gas in Europe, namely the Netherlands and Norway. The UK is a small net importer, and the other large economies are all large net importers. Imports come from Russia, and from North Africa, through pipelines into southern Europe and in the form of LNG into northern Europe.

Price consequences
The combination of long term purchase contracts by the gas companies and their monopolistic position has meant that gas prices in Europe tend to be much less volatile than in the US, though they also tend to be higher and limited only to crude oil prices. The gas utilities manage

the balance of supply and demand through the purchase of peak shaving capacity, sometimes in the form of LNG contracts, though some North Sea fields have been developed to be depleted in this fashion, and through the sale of some bulk gas under interruptible contracts.

There is a less direct connection between the end use market and the wellhead price in Europe than in the US. And, from the point of view of the oil companies, the key to the gas industry is the securing of depletion contracts or long term volume contracts with gas utilities. In Norway, for example, there is a gas marketing body which markets Norwegian gas and which then allocates the supply to individual fields. In the Netherlands, the size of the Groningen field (shared by Shell and Exxon) is such that Gasunie lifts gas from other, smaller fields as it becomes available, and Groningen is in effect the swing producer. The UK has now become complicated because it is in transition between the European and the US models, and oil companies are now having to concentrate on marketing their gas, not just on joining a queue to sell to British Gas under field depletion contracts.

Power generation demand

One feature which the UK shares with the rest of Europe is the shift in political attitude with regard to the use of gas for power generation. Environmental concerns about coal burning and the introduction of the highly efficient combined cycle gas turbine (CCGT) have initiated a 'dash for gas' which has the potential to affect profoundly future European gas market balances. A second large unknown is the reliability and ultimate price of Russian gas.

Oil company involvement

Other than the group of European companies which are involved in potential Nigerian NGL exports to northern Europe (Shell, Agip and Elf), most oil company involvement in the European gas industry is through their equity interests in gas reserves in the Netherlands, Norway and the UK. For the first two, the driving factor is the prospects for their exports into France and Germany. The UK is a more independent market in which the rate at which gas displaces oil (and ultimately nuclear electricity) in the power generation market will probably determine future gas prices.

Erosion of the role of the gas companies should reduce supply margins on average. Since the balance of power lies more with the small number of producers, rather than with the very large number of customers, it may well be that more of the economic rent associated with gas reserves will tend to accrue upstream, to the oil company. The price of gas to the customer has risen against oil so that, in most cases, it is almost equivalent.

Interestingly, the price of gas to power generators has risen in some cases to more than the equivalent cost of oil. This reflects the fact that CCGTs are more efficient than other fossil fuel sources of electricity, and that the market strength of the oil companies is such that they are able to claim some of the margin in their wellhead prices. These points suggest that in future it may be changes to the margin structure of the industry, rather than merely the end user price to the customer, which will affect the wellhead price which oil companies in Europe obtain for their gas.

The Far East

Japanese LNG

The Far Eastern gas market is completely different again. In its contractual structure it is closer to that of Europe, with gas traded under long term supply contracts, with prices linked to spot oil prices. The difference lies in the fact that the main customer for gas in the Far East is Japan, and that gas exports to Japan all take the form of LNG.

Gas in the Far East is mainly produced in Indonesia, in Australia and in Malaysia, though there are smaller amounts produced for internal consumption elsewhere. All three of these countries have LNG contracts with Japanese buying consortia. In some cases, such as the Australian North West shelf, the oil companies are the marketers. In Indonesia, which is the largest supplier, the vendor is Pertamina, the Indonesian state oil company, and the oil companies simply receive their share of net cash flows for production under their production sharing contracts.

Importance of contracts

The value of gas in the Far East depends much more than in Europe and the US on the likelihood of its finding a buyer, and unsold gas far from consumer markets with little prospect of justifying an LNG export contract to Japan is of only notional value. Even in Japan, however, gas has obtained only a small share of the overall energy market, and this combined with environmental pressure seems certain to guarantee fairly rapid demand growth. Moreover, the smaller emerging economies of the Far East are gradually installing an infrastructure to permit them to burn domestically produced supplies, as in Thailand, or to handle LNG imports, as in Taiwan.

Commercial implications

Japanese gas imports are closely linked in price to the cost of oil imports. The netback to the company is dependent on transport and liquefaction costs. Its relatively high sales price and the prospect of a rapidly expanding market has made the Far Eastern LNG business highly attractive to companies which are prepared to undertake large capital investments which pay back over the long term, notably Shell and Total.

One virtually untapped source of LNG is the Middle East, which currently uses gas for domestic power generation but does not export. The first large-scale LNG project to come on stream in the Middle East will be Qatar's North Field in 1998, which will also supply Japan. The capital, technology and marketing skills which LNG projects require probably imply that there is a greater prospect of Western companies being allowed to obtain an equity-type return from them than from oil development in most of the Middle East.

Conclusions

The combination of a move in Europe to direct marketing on the American model and the increasing importance of LNG, rather than gas delivered by pipeline, might be thought to make it likely that the natural gas market would become increasingly like the oil market, with a shift to spot supplies and global prices. This is probably unlikely to occur. The scale of the investment in these projects is such that capacity is sold first and developed afterwards, while the opposite happens to oil.

In addition, the changing structure of the downstream end of the gas business, especially in Europe, implies changing opportunities and risks for the oil companies, even against the background of stable average customer prices. The message for the investment analyst is very clear, however. Evaluation of an oil company's gas reserves requires an understanding of the gas market into which it is selling, as well as the technical and fiscal knowledge required for evaluation of upstream oil assets. As the proportion of the value of oil company asset portfolios which is represented by gas increases, this understanding will become increasingly important.

CHAPTER
THREE

Oil industry accounts

ACCOUNTS AND ECONOMICS

Economics forecasts the future

There is an inevitable tension between accounting and economics. The evaluation of a company is an exercise in applied economics. It involves the creation of forecasts, and the use of some methodology for translation of those forecasts into present value. Whether or not a system of discounted cash flow is explicitly used, it is axiomatic to this process that the value of an asset is determined not merely by how much revenue it will generate, but also when the revenue will be generated. Moreover, for an economist, once a sum of money has been spent, it becomes irrelevant, except to the extent to which it may impact on the future. Evaluation is solely concerned with the future. (This is not to suggest that history does not influence expectations of the future; clearly, it does.)

Accounts record the past

Historical cost accounting is completely the opposite of the above. Accounts record the past. They do not project the future (though they often do involve assumptions about the future). And, in purely historical cost accounts, there is no adjustment made for the impact of inflation on monetary values. The charge taken for the cost of fixed assets is related to their historical cost, not to their replacement cost (although in the case of the depreciation of oil and gas assets expected future capital expenditures are taken into account, historical expenditures are depreciated without adjustment for inflation).

Finding costs unpredictable

In addition to historical costs, conventional accounting procedures are not particularly well suited to the recording of the performance of companies in extraction industries. Because the cost of finding and developing oil varies dramatically from one oil field to another, the assumption that its depletion charge accurately reflects the cost of replacing a fixed asset is more than usually suspect, even if an adjustment is made for the effect of inflation.

Reports and accounts main source

Despite the caveats outlined previously, it should be made clear that reports and accounts are the most detailed and complete source of information which one normally obtains about a company (other than the US Forms 20F and 10K), and much useful information can be obtained from them. Even in the case of a pure exploration and production company (for which, as we shall see in a later section, it is often possible to produce cash flow models of the individual assets, rather than working from consolidated accounts) discounted cash flow models permit establishment of what is, in effect, only a liquidation value. They calculate what an industry acquirer should be prepared to pay for the assets concerned. They do not necessarily give much indication of what an investor can expect to receive in the form of investment return from the performance of the company over time. One guide to this is how well the management of the company has grown the assets over time, and this requires analysis of track record which can only be done using past company reports and accounts.

OIL COMPANY CONSOLIDATED ACCOUNTS

Introduction

There are a number of excellent texts which explain how historical cost accounts are compiled, and which present techniques for interpreting them (e.g. *Interpreting company reports and accounts,* Holmes, G. and Sugden, A., Woodhead-Faulkner, 1989). We do not propose to undertake this task, but shall instead concentrate on highlighting those elements of accounts which are specific to the oil industry, and on certain accounting conventions of which anyone attempting to evaluate an oil company needs to be aware.

By way of introduction, we have taken the preliminary announcement of full year 1991 results by the Royal Dutch/Shell group, and will comment briefly on the three main accounts presented, profit and loss, balance sheet, and sources and applications of funds. These accounts will permit us to refer to most of the issues which are then examined in detail in later sections of this chapter, many of which relate specifically to exploration and production activities, and of which some are best illustrated with reference to the accounts of the independent exploration and production companies.

As can be seen in Chapter 4, it is often possible to build valuation models for exploration and production companies, which are based on appraisal of individual assets. This is not the case with the larger, integrated companies, for which the analysis which is discussed in Chapter 5 is based more on accounts. The Shell Group has been chosen to illustrate this section, partly because it allows us to mention in passing points to which we return in detail later in this chapter, and also because it represents the type of company for which investment appraisal is inevitably based primarily on the consolidated accounts.

Companies are required to make reference to the accounting policies which they are applying in the notes to the accounts, and in the UK, the Oil Industry Accounting Committee has published four *Statements Of Recommended Practice,* to which an increasing number of UK companies are conforming. We shall approach the issue of oil and gas accounting policies primarily from the standpoint of UK accounts. A later section in this chapter considers some of the more important differences between these and some US and European accounting practices.

Profit and loss

Introduction

The statement of income (Table 3.1) shows the group consolidated profit and loss account for the year 1991, and the historical comparison for 1990. It therefore incorporates all of the operating businesses of the group, which mainly consist of exploration and production, refining and marketing, and chemicals operations.

Table 3.1 *Parent companies' share in group net income, and statement of income, for Royal Dutch/Shell*

Parent companies' share in group net income		*1991*	*Full year 1990*
		per ordinary share	
Royal Dutch	N.fl.	9.67	13.98
US dollar equivalents	$	5.23	7.77
Shell Transport	pence	24.9	39.6
New York share equivalents	$	2.66	4.30
Statement of income			£m
Sales proceeds		74 435	73 937
Sales taxes, excise duties and similar levies		16 346	14 521
Net proceeds		58 089	59 416
Cost of sales		45 968	45 022
Gross profit		12 121	14 394
Selling, distribution and administrative expenses		6 401	6 077
Exploration		768	738
Research and development		472	473
Operating profit of group companies		4 480	7 106
Share of operating profit of associated companies		897	993
Operating profit		5 377	8 099
Interest and other income		802	731
Interest expense		818	866
Currency exchange gains/(losses)		79	(71)
Income before taxation		5 440	7 893
Taxation		3 036	4 197
Income after taxation		2 404	3 696
Minority interests		1	87
Net income for the period		2 403	3 609

Source: Royal Dutch/Shell Group: Preliminary announcement.

Trading profit
Sales proceeds, excise taxes and net proceeds are fairly self explanatory. The cost of sales comprises both raw material and labour costs, and the group depreciation charge. It therefore contains a large non-cash element. Selling, distribution and administrative expenses, and research and development costs are as they would be for normal manufacturing companies, but the exploration charge requires some explanation.

Exploration charge
Oil companies constantly explore for new oil reserves, and the largest part of this cost is associated with the drilling of exploration wells. Most of these are unsuccessful (about nine out of ten worldwide). This cost can be treated in different ways (as shown in the section on exploration accounting), but Shell follows the convention of writing off the cost of unsuccessful exploration expenditure as incurred. It should be noted that the exploration charge shown in the profit and loss account is therefore not the amount spent during the year. It is the amount which the group has decided to write off immediately.

Associates and financial items
The share of operating profit of associated companies, and the interest received and charged items are as they would be for a manufacturing company, though it should be noted that oil companies often capitalise a significant proportion of the interest which they actually pay, where the debt is associated with projects under development.

Currency effects
Currency exchange gains and losses tend to be fairly large items for oil companies, other than those accounting in US dollars. The reason for this is that most oil companies generate the larger part of their cash flows in US dollars (in which crude oil and wholesale oil products are priced), and therefore tend to borrow US dollars, in order to match their financial liabilities with their assets. To the extent that they make currency gains or losses on translation of dollar debt or cash holdings into their accounting currency (in Shell's case, sterling), they may record unrealised currency gains or losses in their profit and loss accounts.

This illustrates, incidentally, an important point in oil equity evaluation. The shares do not, in general, offer direct exposure to the currency in which they are notionally denominated; in Shell's case the operating currency of its businesses is the dollar, though its international operations incur revenues and costs in a wide variety of currencies.

Taxation charge

The tax charge for oil companies usually contains not merely the corporation taxes charged on their profits, but also the special oil production taxes levied by governments in their countries of operation. These are sometimes treated as a cost of production, and therefore appear as a cost of sales, rather than as part of the group taxation charge. In Shell's case they are included in the group taxation charge. Tax charged to the profit and loss account and tax paid can diverge considerably, owing to the size of deferred tax charges.

Industry split

Before turning to the balance sheet, we should comment on the statement of earnings by industry segment (see Table 3.2), which is of far greater use in the practical evaluation of the group. (The business which Shell describes as 'manufacturing, marine and marketing' is equivalent to its refining and marketing operations.) This restates the profit and loss account by business contribution net of tax (some groups show the split pre-tax) and then deducts the net of tax charges applicable to finance and corporate functions. Note that Shell follows the historical cost accounting convention for its main accounts, but also provides additional information on a current cost of supplies basis. The significance of this is as follows. The two margin businesses (manufacturing, marine and marketing, and chemicals) operate in volatile markets, and inevitably carry large inventory. Historical cost accounts include gains or losses on inventory; current cost accounts exclude them.

It is the current cost numbers which should be used in appraisal of the performance of the company. Since stock gains and losses are a function of volatility, they are by their nature non-recurring items (though it is true that at times of sharp oil price movement, time lags in passing these through to the product market tend to distort current cost margins as well). In addition, stock price increases or decreases simply have the effect of increasing, or decreasing, the replacement cost of inventory. So the other side of the coin to a large oil price increase, which boosts profits, is a large increase in working capital owing to higher cost of inventory. In cash terms, the company is no better off.

Table 3.2 *Royal Dutch/Shell earnings by segment*

	1991	Full year 1990
Earnings by industry segment		£m
Oil and gas:		
Exploration and production:		
World outside US	1436	1589
US	3	234
	1439	1823
Manufacturing, marine and marketing:		
World outside US	1263	1691
US	(99)	53
	1164	1744
Chemicals:		
World outside US	(60)	364
US	83	155
	23	519
Other industry segments:		
Coal	(33)	36
Metals	8	97
Other	(12)	(2)
Earnings from operations	2589	4217
Corporate items:		
Currency exchange gains/(losses)	(33)	(210)
Other – including interest income/expense and taxation	(152)	(311)
Total corporate items	(185)	(521)
Minority interests	(1)	(87)
Net income for the period	2403	3609
Earnings on an estimated current cost of supplies basis		
Oil and gas:		
Manufacturing, marine and marketing:		
World outside US	1718	1108
US	(97)	43
Chemicals:		
World outside US	(27)	362
US	83	155
Earnings on an estimated current cost of supplies basis	2893	3014

Source: Royal Dutch/Shell Group: Preliminary announcement.

Balance sheet

Fixed assets

Taking each of the items of the balance sheet (see Table 3.3) in turn, the fixed assets mainly comprise the property, plant and machinery in the

Table 3.3 *Royal Dutch/Shell: summarised statement of assets and liabilities*

	December 31 1991	*December 31 1990*
	£m (translated at end rates of exchange)	
Fixed assets:		
Tangible fixed assets	34 120	31 898
Investments	4 250	4 064
	38 370	35 962
Current assets:		
Inventories	4 491	5 222
Accounts receivable	8 871	9 551
Cash, short-term deposits and securities	4 703	4 411
	18 065	19 184
Current liabilities:		
Short-term debt	3 006	3 471
Accounts payable and accrued liabilities	7 367	7 848
Taxes payable	3 067	3 237
Dividends payable to parent companies	1 194	1 142
	14 634	15 698
Net current assets	3 431	3 486
Total assets less current liabilities	41 801	39 448
Long-term liabilities:		
Long-term debt and capitalised lease obligations	3 587	2 322
Other	488	446
	4 075	2 768
Provisions:		
Deferred taxation	6 078	6 246
Other	2 142	1 941
	8 220	8 187
Minority interests	626	639
Net assets	28 880	27 854

Source: Royal Dutch/Shell Group: Preliminary announcement.

downstream (refining, marketing and chemicals) businesses, as well as the capitalised values of the upstream (exploration and production) properties. In the case of exploration and production assets it is important to note that the balance sheet value of the assets is based on historical expenditures, and bears little relationship to their commercial value (though the values are subject to a ceiling test, described later).

For the exploration and production businesses, including those inside integrated oil companies, it is the relationship between the historical cost of the tangible fixed assets in the balance sheet and their commercial value which represents the success or otherwise of the company in creating value through its operating activities. The notes to the balance sheet are therefore of great importance to the analyst, especially in conjunction with the details provided in the statement of oil and gas reserves (discussed later in this chapter). Evaluation of exploration and production operations essentially comprises a revaluation of their tangible fixed assets.

Working capital

Most of the working capital is attributable to the downstream businesses. Exploration and production businesses carry virtually no inventory, as most oil is transferred fairly promptly after production at the well head. It will be noted that the sharp decline in the value of Shell's inventory over the year 1991 is in line with the stock loss represented by the difference between the historical cost earnings from the refining and marketing business, and the current cost earnings.

Capital employed

Shell's capital employed consisted of long term debt, provisions and equity, of which the notable feature is the size of the provisions. Oil companies often have high deferred tax charges owing to the structure of upstream oil taxation, which often permits exploration and capital development costs to be depreciated for tax purposes in excess of the accounting depreciation. In calculations of return on capital, it is conventional to net off provisions, using only debt and equity to derive capital employed.

Cash flow

Link with economics

In addition to being the link, in accounting terms, between the profit and loss account and the balance sheet, it is the cash flow statement (as shown in Table 3.4) which provides the link between the accounting of a company and its economic evaluation through, for example, discounted cash flow techniques. In pure economic theory, the value of a company to its investors is the discounted net present value of its future free cash flows from operations, that is to say, its cash flows from operations after tax, less the capital expenditure required to generate those cash flows.

Exploration companies
As we shall see in Chapter 4, in the case of exploration and production companies with not too many individual projects, it is possible to consoli-

Table 3.4 *Royal Dutch/Shell: summarised statement of cash flows*

	1991	*Full year 1990*
	£m (translated at average rates of exchange)	
Cash flow provided by operating activities:		
Net income for the period	2403	3609
Depreciation, depletion and amortisation	3644	3286
Decrease/(increase) in net workings capital	907	(293)
Associated companies: dividends more/(less) than net income	(75)	(171)
Deferred taxation and other provisions	(117)	225
Other	81	28
Cash flow provided by operating activities	6843	6684
Cash flow used in investing activities:		
Capital expenditure	(5722)	(5201)
Proceeds from sale of assets	540	136
Net investments: associated companies and other	(94)	(114)
Cash flow used in investing activities	(5276)	(5179)
Cash flow provided by/(used in) financing activities:		
Net increase/(decrease) in long term debt	976	125
Net increase/(decrease) in short term debt	(362)	1044
Dividends paid: to parent companies	(1991)	(1913)
to minority interests	(34)	(37)
Cash flow provided by/(used in) financing activities	(1411)	(781)
Currency translation differences relating to cash and cash equivalents	136	(376)
Increase/(decrease) in cash and cash equivalents	292	348

Source: Royal Dutch/Shell Group: Preliminary announcement.

date cash flow models built up on an asset by asset basis. The flow of funds which this procedure generates is in effect a projection of the net stream of cash which will be generated by its existing assets and be available to the company to pay down debt, to distribute to shareholders, or to plough back into new projects. It is a liquidation value.

Integrated companies
In the case of the larger, more complicated companies such as Shell, we discuss evaluation procedures in Chapter 5, but since evaluation of individual assets is virtually impossible, one conventional approach is to establish the surplus which is left over, again for growth or for distribution, after adequate provision has been made for maintenance of the existing assets. The difference between this estimate and the cash generated by operations is the addition to the economic value which has been achieved by a company over a given period.

Stock effects
Before turning to directly evaluative methods, we shall look rather more closely at some of the accounting conventions used by oil companies, in order to facilitate their interpretation. In this context, it is notable that the effect of the sharp fall in oil prices during 1991 appears again in Shell's cash flow statement. It depressed the historical cost profit and loss account. It was reflected in a fall in the value of inventory between end 1990 and end 1991. And it appears again as a source of funds from working capital reduction in the sources and applications of funds. The following section opens with a look at the issue of treatment of inventory.

OIL INDUSTRY ACCOUNTING CONVENTIONS

Inventories

LIFO and FIFO
Although the details vary, there are basically two ways in which to account for oil inventory. These can be most simply categorised as First-In-First-Out (FIFO) and Last-In-First-Out (LIFO). They approximate to the historical cost and current cost of supplies methods used in the Shell accounts. Table 3.5 shows a simplified calculation of profit per barrel

refined, illustrating the basic difference between the two conventions, as far as the profit and loss account is concerned. It assumes a two month inventory cycle for the refiner. On the FIFO basis, a product barrel sold in March for $21 has debited against it the cost of a barrel of crude oil bought in January for $18, with a resulting gross profit of $3 per barrel. On the LIFO basis, the same product barrel has debited against it the cost of a replacement barrel of crude oil in March, with a resulting margin of $1 per barrel.

Implications for accounts

Fairly clearly, in the event that LIFO margins remain constant, FIFO profits will increase when oil prices are rising, and will decrease when prices are falling, but the underlying trading position has not changed. The choice of inventory accounting convention also has implications for balance sheets and sources and applications of funds statements. For FIFO accounters, oil price increases or decreases result in increase or decrease in the balance sheet carrying value of inventory, as it is 'old' barrels which are debited against sales and 'new' barrels which are carried in the balance sheet. For a LIFO accounter, if the level of inventory remains stable then there is never any change in its carrying value. It is only the 'old' barrels which are carried in the balance sheet. LIFO accounters only take stock gains or losses when they draw down reserves, and a current sale is debited with a past purchase. This means that LIFO balance sheets often understate actual capital employed.

Hybrid accounts

In addition to those favouring either LIFO or FIFO, there are in addition some hybrid companies, such as Elf Aquitaine. Elf states its inventory in the balance sheet on a FIFO basis, while its profit and loss account is effectively LIFO. This is because it provides, each year, for the effect of movement in oil prices on the replacement cost of inventory. Thus, it does not take the effect of price movements on inventory through the

Table 3.5 *Notional refined barrel (\$/bbl)*

	January	*March*
Crude oil price	18	20
Product barrel	–	21

FIFO margin = \$3/bbl
LIFO margin = \$1/bbl

profit and loss, and its balance sheet contains a provision which offsets the increase or decrease in value of inventory.

Market values

Whatever system of accounting is used, prudence dictates that the carrying value of inventory in the balance sheet should not exceed its market value. In an oil price crash, such as in 1986, even LIFO accounters may report stock losses, owing to this adjustment.

Cash flows

Finally, as has already been noted, the choice of accounting treatment for inventory should have no significant impact on stated cash flow from operations. Stock gains on inventory accounted for under the FIFO convention will be offset by an outflow of cash to fund increased working capital (unless inventory has been reduced over the period). Also, stock losses due to falls in oil prices should be offset by release of cash from reduction in working capital, as the Shell accounts illustrated.

Exploration

Introduction

If the primary accounting distinction between refining and marketing businesses is whether or not they account using something akin to the LIFO or FIFO convention, the main item which distinguishes the accounting treatment of exploration and production businesses is whether they use the full cost or the successful efforts convention for expensing exploration activity.

Successful efforts

Under the successful efforts system, exploration expenditure is immediately categorised as either having been unsuccessful or as justifying being held in suspense, depending upon whether the result from the well drilled justifies further appraisal, or makes commercial development a probability. Once the reserves are deemed commercial then a cost pool is created into which all development and abandonment costs should be capitalised, and against which financing costs related directly to capital expenditure may be capitalised.

> *All expenditure carried within each field should be depreciated on a unit-of-production basis by reference to quantities . . . The cost element of the unit-of-production calculation should be the costs incurred to date together with the estimated future development costs of obtaining access to all the reserves included in the unit-of-production calculation.* (*SORP 2*)

Full cost

Under the full cost accounting system, the full cost of all exploration, appraisal and development expenditure is capitalised within an appropriate cost pool, 'except that certain exploration and appraisal costs may be held in suspense outside cost pools pending determination' (*SORP 2*). The depletion of the capitalised value in the pool is calculated not on an asset by asset basis, but by applying the unit of production basis to the entire cost pool. The basis under which cost pools are established is at the discretion of the company, but should be disclosed as an accounting policy.

Ceiling tests

In both cases, companies should apply ceiling tests

> *to determine whether the net book amount of expenditure within each cost pool less any provisions for abandonment costs and deferred production or revenue related taxes is covered by the anticipated future net revenue from oil and gas reserves attributable to the company's interest in that pool.* (*SORP 2*)

In neither case are corporation taxes or financing costs deducted in the calculation. Prices and costs are unescalated, and no discount rate is applied to future cash flows.

Accounting implications

The effect of the accounting convention used on published accounts tends to be as follows. The balance sheet of the full cost accounter usually carries a higher item for fixed assets than that of a successful efforts company, because it will include items of capitalised unsuccessful exploration expenditure, which the successful efforts company will have written off. Its gearing ratio will therefore be correspondingly lower.

The profit and loss account will usually show a higher profit prior to exploration charges for the successful efforts company for the same reason: it has less to deplete. The successful efforts company, however, has to write off the unsuccessful exploration expenditure which it has undertaken in that particular year, and if it has had a large or unsuccessful exploration programme, this may more than offset the lower depletion charge.

There will be no effect on stated cash flows, so long as the successful efforts company writes back its exploration charge as a source of funds, and includes the full amount spent on exploration, whether capitalised or not, as capital expenditure. If it does not, then its cash flow from operations will be understated as against a full cost accounter. The net cash flow, of course, can never be affected by any change of accounting convention: it is what actually flowed into or out of the company over the year.

An illustration of the effect of restating from one accounting convention to another is offered by LASMO's choice, in 1991, to switch from the full cost to the successful efforts convention. The cost of sales in the profit and loss account for 1990 (shown in Table 3.6) has been reduced considerably by the reduction in depletion charge. But, because of the extent of LASMO's exploration programme, the write-off of unsuccessful exploration expenditure actually exceeded the reduction in depletion, with the effect of reducing sharply the level of operating profit stated. This is despite LASMO's decision to credit the offset against petroleum revenue tax attributable to exploration expenditure against the exploration charge, rather than treating it as a credit against the tax charge, which would have been a more usual practice.

Table 3.6 *LASMO 1990 profit and loss account*

			£m
	1990 (original)	*1990 (restated)*	
Turnover	358.1		332.6
Cost of sales	(253.3)		(231.2)
Gross profit	104.8		101.4
Exploration charge		(82.6)	
PRT relief		16.1	
			(66.5)
Administrative expenses	(16.9)		(15.3)
Operating profit	87.9		19.6

Source: LASMO.

Reflection of maturity

Clearly, in the early years of a company's life, use of the full cost accounting system will result in a stronger balance sheet, and avoid the almost inevitable accumulation of losses which will be generated by a successful efforts company which is exploring actively, but which has little production. The problem with full cost accounting as the company expands is that the exploration programme is capitalised at its gross cost, despite the fact that companies are often able to offset much of the cost of drilling against their tax liabilities, and take their exploration decisions on the basis of the net cost of undertaking the expenditure. This can result in overstatement of the value of the cost pools, and, when the company reaches maturity, in an excessive depletion charge which it becomes ever harder to offset with income.

Deferred tax

Introduction

Deferred tax arises as an issue for all companies. The charge to the profit and loss account for a given period represents the liability to taxation which has accrued as a result of the taxable surplus which the company has generated during the accounting period, whereas the amount actually paid may be somewhat different from this.

Tax allowances

The simplest source of timing difference is attributable to payment of tax in the year following accrual of a liability. Rather larger and more significant timing differences occur in cases in which the depreciation of assets used in determination of taxable profit (the figure on which taxation is charged) is faster than that used in determination of accounting profit (the figure which appears as profit before taxation in a company's profit and loss account). In these cases, companies usually apply a deferred tax charge to the profit and loss account, in order to charge against the profits in a given accounting period the taxation to which that profit will give rise, even if it will not crystallise for some time.

Cost recovery

The simplest example of companies deferring tax liability in the oil industry is the practice of allowing full cost recovery of capital expended on the development of a project, prior to liability being incurred. An alternative example is the practice of allowing exploration expenditure to be written off for tax purposes against the taxable profits arising from revenues generated from producing fields. Deferred tax arises where the timing of tax allowances is different from the timing of the depreciation of costs for normal accounting purposes.

Corporation tax effects

As far as corporation tax is concerned, the main difference between companies is over whether they fully defer for all tax liability, whether or not they can foresee its crystallising – as is required in the US – or whether they provide only for that which they can foresee crystallising, as in the UK. The difference between the two is illustrated by a US GAAP return from BP (which for the purposes of its main accounts follows the UK accounting convention). Most of the continental European companies produce accounts under the US GAAP standard, or something approximating to it. National accounting standards vary widely across Europe, however, and in many cases under domestic accounting standards, the provisioning which is permitted for tax purposes is reproduced in the stated accounts, which has the effect of severely understating earnings. Norsk Hydro is one company which reports under both local (Norwegian) and US GAAP conventions, and the 1991 profit and loss accounts under the two standards shown in Table 3.7 highlight the differences of treatment.

Upstream tax effects

When it comes to special upstream taxes, the two most significant are those in the UK and in Norway. For accounting purposes, UK petroleum revenue tax causes the greatest complications, since it is levied not on the company but on the field, and the calculation of taxable profit for the field is materially different from the calculation of accounting profit accruing from equity in the field for the oil company (or companies) involved. For a description of different types of upstream taxation see Chapter 4, which covers the valuation of exploration and production companies.

Life of field and timing differences

There are basically two ways in which to account for petroleum revenue tax: 'life of field' accounting, and 'timing difference' accounting. In life of field accounting, the entire remaining liability to the tax is calculated for each field, and the company then provides for it on a unit of production basis. In timing difference accounting, the procedure followed is much closer to that for corporation tax:

> *Deferred petroleum revenue tax is provided on those timing differences which arise when exploration and development costs are written off, or depreciated for accounting purposes, in periods different from those in which the expenditure is allowable for petroleum revenue tax.* (ARTHUR ANDERSEN, *FINANCIAL REPORTING BY THE EUROPEAN OIL INDUSTRY*, 1991)

As with corporation tax, the timing difference approach results in a choice between full deferral, and deferring only for those liabilities which are foreseen to crystallise. It also allows for some quite substantial differences between treatment of such items as uplift, oil allowance and safeguard.

Table 3.7 *Norsk Hydro consolidated income statements reported under (a) US GAAP and (b) Norwegian accounting conventions*

	Year ended 31 December		
Amounts in NOK million except per share data	*1991*	*1990*	*1989*
Operating revenues	61 423	60 972	66 329
Operating costs and expenses:			
Raw materials and energy costs	37 890	33 066	38 370
Payroll and related costs	9 056	8 388	9 080
Depreciation, depletion and amortisation	5 193	4 673	4 832
Other	5 459	7 903	7 048
Restructuring charges	2 900	200	–
Total operating costs and expenses	60 498	54 230	59 330
Operating income	925	6 742	6 999
Equity in net income of non-consolidated investees	205	167	162
Financial income (expense), net	(1 207)	(1 206)	(1 539)
Other income (expense), net	514	37	195
Income before taxes and minority interest	437	5 740	5 427
Current income tax expense	(737)	(944)	(989)
Deferred income tax expense	(139)	(1 804)	(1 608)
Minority interest	(59)	(91)	(143)
Net income (loss)	(498)	2 901	2 687
Earnings (loss) per share	(2.40)	14.10	13.10

(a)

Amounts in NOK million	*Year ended 31 December* *1991*	*1990*	*1989*
Operating revenues	61 423	60 972	66 329
Operating costs and expenses:			
Raw materials and energy costs	38 134	33 504	39 438
Payroll and related costs	9 423	8 845	9 398
Depreciation, depletion and amortisation	4 562	4 147	4 421
Other	5 794	7 888	7 226
Restructuring charges	2 900	200	–
Total operating costs and expenses	60 813	54 584	60 483
Operating income	610	6 388	5 846
Equity in net income of non-consolidated investees	205	167	162
Financial income (expense), net	(1 680)	(2 199)	(2 202)
Other income (expense), net	514	37	195
Income (loss) before minority interest, appropriations and taxes	(351)	4 393	3 611
Current income tax expense	(737)	(944)	(989)
Tax related appropriations	(1 022)	(1 314)	(974)
Minority interest	(59)	(91)	(143)
Net income (loss) stated in accordance with Norwegian accounting principles	(2 169)	2 044	1 505

(b)

Source: Norsk Hydro.

Company size differential

As with accounting for exploration expenditure, there is a tendency for the procedures used by larger and smaller companies to diverge. The larger, international companies tend to use the timing difference method. Many of the UK exploration companies use the life of field method. For small companies in particular, the use of the life of field method avoids sharp changes in tax charge in the event that, for example, the field moves out of safeguard relief.

Oil price fluctuations

Life of field computations depend on the oil price assumption used in calculating the entire liability over the remaining life of a field, and this can require some dramatic adjustments when oil price expectations change. The example of Enterprise Oil's 1986 earnings was a case in point. The collapse in the oil price during that year resulted in a situation in which previously deferred amounts with respect to petroleum revenue tax liability exceeded that which the company could ever foresee paying at year end oil prices. As a result, it was able to write back previously deferred tax through its 1986 profit and loss account; this was a material factor in the company's ability to cover a held dividend for the year, at a time when most other exploration and production companies were forced to reduce their payouts. A similar situation arose in 1993 when the rate of petroleum revenue tax was reduced.

Oil and gas reserves

Introduction

The accounting conventions discussed so far are fairly clear cut. A company is either a LIFO or a FIFO accounter of inventory, a successful efforts or a full cost accounter of exploration expenditure, or a timing difference or a unit of production accounter of deferred tax. As the reader will have noticed, there are substantial differences of treatment permissible in many cases, but the broad impact of the chosen convention on a company's accounts is fairly straightforward. This is not the case for the treatment of oil and gas reserves.

Reserve depletion

In both the discussion of depletion of exploration and development costs and of the unit of production method of deferring for petroleum revenue tax, mention was made of the principle of depreciating a liability proportionately with the rate at which reserves are produced, whether on a field by field basis or within a cost pool. Clearly, to do this it is necessary to establish, for accounting purposes, what the reserve is, against which the overall charge is to be depleted. This overlaps with the issue of disclosure of reserves in the accounts of an oil company, though it is not the case that the reserves allocated to cost pools for accounting purposes are identical with the reserves shown in the notes to the accounts.

Chapter 4 of this book, on the evaluation of exploration and production companies, concerns itself with definition of oil and gas reserves, geologically and commercially, and their evaluation. This section is concerned with accounting, and we shall therefore restrict ourselves to what appears in oil company accounts, and how it affects the main accounts items.

Accounting definitions

Accounting definitions of reserves combine two concepts explored more fully in Chapter 4, namely, the geological issue of what is recoverable, and the commercial issue of what it is reasonable to expect is commercially recoverable. *SORP 1* gives the following definitions regarding oil and gas reserves:

- *Proved reserves* are described as 'the estimated quantities of crude oil, natural gas and natural gas liquids which geological and engineering data demonstrate with reasonable certainty to be recoverable in future years from known reservoirs under existing economic and operating conditions, that is, prices and costs as at the date the estimate is made'.
- *Probable reserves* are 'those which at present cannot be regarded as proved but which are estimated to have better than a 50% chance of being technically and economically producible'.
- *Possible reserves* are 'those reserves which at present cannot be regarded as probable but are estimated to have a significant but less than 50% chance of being technically and economically producible'.

Recommended practice

SORP 1 goes on to recommend that companies should report the net quantities of their proved reserves, whether developed or undeveloped, as at the beginning and end of each year, reconciling movements for the effects of revisions, purchases, extensions and discoveries, sales and production. It also recommends that

> *Where unit-of-production methods are used in calculating provisions for depreciation, depletion and amortisation, decommissioning or taxation, the reserve categories (proved, probable or possible) and the proportions of each category used be given together with a description of the related cost centres.*

Unfortunately, there remains considerable diversity between oil companies as to the categories of reserve used for the purposes of depreciation.

Production sharing contracts

Before moving on to statements of reserves, a particular problem should be noted with regard to production sharing contracts. The simplest form of fiscal system for exploration and production companies is that under which they are awarded equity ownership of the oil and gas reserves in a particular licence area, and then pay tax on the profit (whichever way defined), which accrues as a result of production from reserves in the area. This may, as in the US, take the form of simple corporation tax, with little additional liability, or, as in the UK and Norway, substantial additional taxes may be applied. As we shall see in Chapter 4, one of the complicating factors in the UK is that the calculation of taxable liability for petroleum revenue tax is different from the basis used for corporation tax.

There are, however, other systems of taxation of exploration and production activities which vary fundamentally from this basic model of equity and profit tax, and they create accounting problems with regard to reserves and depletion (which is why they must be addressed in passing in this section). In many parts of the world, e.g. Indonesia and many of the Middle Eastern countries, oil companies are not awarded any equity in oil and gas reserves. Instead, they are offered a production sharing contract, under which they are entitled to a proportion of the cash flows attributable to production from the licence area. Typically, the entitlement is high until the company has recovered the capital sunk during exploration and development, and is then reduced once the project has paid back.

Entitlement oil

The problems which production sharing contracts raise in accounting are clear. Strictly, a company in this situation has no reserves over which to deplete its capitalised expenditure. Moreover, the proportion of cash flow which will accrue to it from production of the reserves will vary. At high oil prices, the project will pay back faster, and the company's entitlement to a proportion of the cash flow from the project will fall faster than it would under low oil prices, in which case payback would be deferred. Moreover, if reserves are stated in the accounts on the basis of the company's estimate of entitlement, rather than its working interest in the licence, it will appear to have found a significantly smaller number of barrels than it has actually found. Thus, to a reader of the accounts, its exploration performance will look poor. If the working interest is used, the revenues per barrel of oil equivalent produced will seem unaccountably low.

An example of the confusion which this can cause is provided by Ultramar, which (prior to LASMO's successful bid in 1991) used to account for its Indonesian exploration and production assets (mainly natural gas) on the basis of its entitlement interest, but which showed reserve and production data in its report and accounts on the basis of its working interest.

In cases in which the reserves concerned represent a significant proportion of those attributable to a company, the situation is clarified considerably if that company accounts on the basis of estimated entitlement, but shows reserves stated on both entitlement and working interest basis.

Reserve replacement

As mentioned earlier, oil companies now almost all include figures for reserve movement over the year as a note to the accounts, and this provides perhaps the most useful measure (in conjunction with the exploration cost) of their performance over the year. All companies engaged in resource extraction are involved in a constant struggle against exhaustion of their reserves.

One of the most important elements of their performance is how successfully they have replaced their reserves through exploration effort, and how expensively they have done this (the calculation should always be made over a run of years, perhaps three to five, since individual figures may be misleading). These figures alone do not illustrate whether or not they have added to their value, which requires the kind of appraisal described in the next section of this book. But it does permit a rapid assessment of sustainability, and of comparative performance.

Disclosures differ

Unfortunately, comparison between companies is not facilitated by the fact that they do not all reveal reserves by similar categories. Most large companies display their proved reserves by geographical area. Many smaller exploration and production companies display total figures for proved and probable reserves. Only rarely are the proved, probable and possible categories shown.

For the purposes of illustration, BP's reserve disclosure statement from the firm's 1991 annual report and accounts is shown in Table 3.8. As is customary for the major companies, BP's disclosure relates to proved reserves only. It displays the movement from the closing balance at end 1990 to end 1991, breaking down the factors behind the movement between revisions to previously estimated reserves, extensions and additions attributable to exploration activity, purchases and sales, and production over the course of the year.

Problem of interpretation

Reserve assessment is based on information which is subject to change, hence the inevitability of revisions. These mainly arise from increase in estimates of ultimately recoverable reserves from fields which have been in production for a period. Such is the caution in the industry that revisions often effectively double the reserves of a field as originally booked after exploration and appraisal.

Assessment of performance tends to concentrate on the balance over time between extensions and discoveries on the one hand and pro-

Table 3.8 *BP supplementary oil and gas information (unaudited)*

Estimated net proved reserves of crude oil	*United Kingdom*	*Rest of Europe*	*US*	*Rest of World*	*Total*
					(mmbbl)
1991					
Subsidiary undertakings					
At January 1					
Developed	806	209	1 622	108	2 745
Undeveloped	773	80	851	204	1 908
	1 579	289	2 473	312	4 653
Changes in year attributable to					
Revisions of previous estimates	(27)	2	48	(94)	(71)
Redetermination of interests	(6)	–	–	–	(6)
Sales of reserves-in-place	(24)	–	(27)	–	(51)
Extensions, discoveries and other additions	91	–	136	9	236
Improved recovery	–	–	118	2	120
Production	(131)	(30)	(269)	(14)	(444)
	(97)	(28)	6	(97)	(216)
At December 31					
Developed	712	194	1 590	91	2 587
Undeveloped	770	67	889	124	1 850
	1 482	261	2 479	215	4 437
Associated undertakings					
Group share					
At January 1					2 077
Net revisions and other changes					(33)
Production					(51)
At December 31					1993
Total group and BP share of associated undertakings					6430

Estimated net proved reserves of natural gas	*United Kingdom*	*Rest of Europe*	*US*	*Rest of World*	*Total*
					(bcf)
1991					
At January 1					
Developed	2 429	78	2 401	1 629	6 537
Undeveloped	2 717	6	639	1 124	4 486
	5 146	84	3 040	2 753	11 023
Changes in year attributable to					
Revisions of previous estimates	(185)	10	24	(21)	(172)
Redetermination of interests	12	–	(1)	16	27
Purchases (sales) of reserves-in-place	(183)	–	(221)	1	(403)
Extensions, discoveries and other additions	286	–	150	520	956
Improved recovery	–	–	1	1	2
Production	(263)	(10)	(128)	(109)	(510)
	(333)	–	(175)	408	(100)
At December 31					
Developed	1 866	82	2 268	1 711	5 927
Undeveloped	2 947	2	597	1 450	4 996
	4 813	84	2 865	3 161	10 923

Source: BP.

duction on the other, and on the relationship between exploration expense and the level of extensions and discoveries (finding costs per barrel). Comparisons may therefore penalise companies with a more cautious approach to booking reserves.

Accounting for value

Introduction

Mention was made earlier of the publication of accounts of calculated values for oil and gas reserves. In theory, an alternative to successful efforts or full cost accounting is offered by the system of value accounting for reserves. Under this system, reserves are annually revalued on the basis of an appraised value, in effect replacing a historical cost valuation of fixed assets with a calculated value.

Subjectivity

Despite the fact that the system of value accounting for reserves is the only way in which reports and accounts could ever accurately reflect the value of exploration and production companies, it is almost never used. The main reason for this is that reserve evaluation, as we shall see in the next chapter, is highly subjective. There are circumstances under which oil company accounts do include such evaluations, however, and these include US listed equities (where such evaluation is required in the notes to the accounts) and elsewhere, e.g. in listing particulars for companies coming to the equity market in the UK.

Technical discussion of these evaluations fits more naturally into the next chapter of this book, but we shall note some points on the accounting here. Where these figures are published in company annual reports and accounts, they represent important additional information, and it is therefore necessary to understand how it should be interpreted.

US SEC requirements

The US requirement has the benefit of clarity, even if the figure derived is somewhat artificial. The note from BP's 1991 Form 20F (Table 3.9) presents this statement under US accounting conventions. All oil and gas asset evaluations use discounted cash flow techniques to establish the net present value of future cash flow from operations for that asset, on the assumption of a given future oil price trend, cost inflation, and using a given discount rate. These are, as it were, external assumptions.

The technical issues surrounding valuations tend to concentrate on the internal assumptions, such as reserve category, depletion rates, real costs, and so on. Acreage and unproven reserves extend the problem of evaluation far beyond a simple cash flow model, as we shall see.

The US requirement is for value to be calculated only for proved reserves (developed and undeveloped), and for the cash flows to be modelled assuming unchanged nominal year end prices and costs. A discount rate of 10% is prescribed. This has the advantage of simplicity and universality, and the disadvantage that it gives a highly misleading representation of the actual market value of the fixed assets of an exploration company.

UK disclosures

In the UK, calculations of value are only required in admission to listing particulars, but without clear guidelines of valuation method, but they do often appear in other corporate documents, such as defence docu-

Table 3.9 *BP's presentation of its oil and gas asset evaluations under US conventions*

Future net cash flows have been prepared on the basis of certain assumptions which may or may not be realised. These include the timing of future production, the estimation of crude oil and natural gas reserves and the application of year end oil and gas prices and exchange rates. Furthermore, both reserve estimates and production forecasts are subject to revision as further technical information becomes available and economic conditions change. BP cautions against relying on the information presented because of the highly arbitrary nature of assumptions on which it is based and its lack of comparability with the historical cost information presented in the financial statements.

	United Kingdom	*Rest of Europe*	*US*	*Rest of World*	*Total*
					(£m)
At December 31 1991					
Future cash inflows	22 600	2 400	13 800	5 700	44 500
Future production and development costs	15 900	1 600	11 300	2 500	31 300
Future taxation	1 900	200	600	900	3 600
Future net cash flows	4 800	600	1 900	2 300	9 600
10% annual discount	1 800	100	600	1 300	3 800
Standardised measure of discounted future net cash flows	3 000	500	1 300	1 000	5 800

Source: BP.

ments and rights issue documents. In these cases, an external authority is generally employed by a company to produce an independent evaluation of its assets. This will often include other categories of reserves than merely the proved, though even then it is not usual to value exploration acreage. External assumptions, such as oil price trends, cost inflation and discount rates are at the discretion of the engineer or consultant producing the evaluation. Moreover, no value is generally applied to the taxation synergies which may accrue to the company from, for example, its exploration programme, or its means of financing.

Abandonment

Introduction

Another accounting convention which is specific to oil companies, and which can have a dramatic impact on the apparent profitability or otherwise of a company's exploration and production operations, is the question of provision for abandonment. This issue principally arises in the case of offshore platforms, for which the cost of site restoration (depending on the extent of the government's requirements in this matter) could be enormous, and the liability at the end of the field's life is a significant factor in its value.

Recommended practice

SORP 3 covers the issue of abandonment, and it recommends that provisions should be made on a field by field basis, by both full cost and successful efforts accounters, using estimated costs at the balance sheet date. It states that 'the charge to income should not be discounted for time'. Where there is a change in estimate, the estimated total cost less the cumulative amount already provided should be allocated to remaining reserves (the 'prospective' method).

Assessing liability

Apart from the question of how much restoration will be required of them, an additional difficulty for companies is to assess whether their facilities will continue to serve a function after depletion of the reserves which they were originally intended to produce. A good example of this instance is the Frigg gas field facility, on the UK/Norway median line. The gas in the Frigg field is largely depleted, but the Frigg to St Fergus gas pipeline will almost certainly be used to transport new contract volumes of Norwegian gas to the UK. This gas will be handled using facilities at the Frigg field. It is therefore perhaps appropriate that Total, in its accounts, should not charge for the abandonment of this facility.

The same consideration applies to platforms which are likely to be used for the development of satellite fields, often using subsea completions hooked into facilities which were originally installed to produce reserves from fields which are now largely depleted. Some scepticism should perhaps, however, be attached to accounts which make no provision whatsoever for abandonment, such as those of Norsk Hydro; the basis being that under the terms of Norwegian concessions, the state has the option of taking ownership of fields and facilities at the end of concession periods, without compensation, but incurring the obligation for abandonment. Equally, however, the state could demand a company undertake restoration work after depletion, or at the end of the concession period, and it is perhaps optimistic to assume that the state will relieve a company of all of its abandonment obligations.

Residual assets

With the increasing use of floating production systems in the UK North Sea, far from facing a liability, producers may find themselves with a valuable asset once a field has been abandoned. This should clearly be taken into account in the evaluation of companies, but will not be reflected in the accounts.

Currency effects

Introduction

Many of the issues regarding accounting treatment of finance for oil companies apply equally to other industries. The main factor which is specific to oil companies is their exposure, at least for exploration and production activities, to a purely US dollar denominated stream of revenue, even if the operations concerned are not geographically situated in the US.

Dollar revenues

Crude oil is priced internationally in US dollars. Even in the case of integrated companies, which may be passing their own oil from their exploration and production subsidiaries to their refining and marketing subsidiaries rather than selling it to third parties, the transfer price used will be in dollars, and it will be the refining business which is exposed to a currency risk.

Dollar debt

Oil companies respond to the situation regarding currency risk exposure by financing their upstream activities through predominantly dollar debt. Their refining and marketing businesses, on the other hand, may well borrow in local currencies (as those are the currencies in which they sell their products in individual domestic markets) to fund their capital expenditure, but then hedge their dollar exposure to the purchase of crude oil as it is incurred through operations.

Currency gains and losses

The main accounting question arises from how to treat gains or losses on debt (or cash balances) which is held in US dollars, but which is then translated into local currency for the purposes of reported accounts. In

> *The carrying party should combine costs, revenues, oil and gas reserves and production attributable to the carried interest with those associated with its own equity interest. Its turnover and cost of sales should include revenues and operating costs, respectively, associated with the carried interest. Unit of production calculations should include the expenditures and oil and gas reserves and production associated with the carried interest, as well as with the equity interest. Consequently, capitalised costs relating to the carried interest should be amortised over the life of the field as a whole, rather than just over the period during which reimbursement occurs.*
>
> *The carried party should not account for carried costs or the related revenues, oil and gas reserves or production from which such costs are recovered.*

The Miller example

In the case of a small oil company with an interest in a large field which it would find difficulty in developing alone, a carry is an alternative to project finance. Whether or not it is more attractive will depend on the attitude of the banks, and of potential carrying parties in the oil industry. An example of this type of arrangement was that made by Saxon Oil Ltd (now part of Enterprise Oil). Saxon's main asset was its interest in the undeveloped Miller field, which was too much for it to fund on its own. Rather than be forced to reduce its interest, it arranged for the operator, Santa Fe, to carry it through the development costs which were attributable to part of its stake. Enterprise describes the arrangement in its report and accounts, but no similar disclosure is made by the carrying party, Santa Fe.

Project finance

The alternative financial arrangement available to a company which is unable (or unwilling) to raise adequate corporate debt is to raise borrowings in the form of non-recourse or limited recourse finance. This is secured directly against the asset concerned, and repayable directly out of the cash flows which the asset generates, and is known as project finance. In this instance, the bank assumes a direct share in the risks associated with the project concerned, and interest charges incurred prior to the field coming on stream are rolled up. The cost of debt tends, therefore, to be higher.

As *SORP 4* points out, while it would appear logical to net off the liabilities under the financing against the owner's capitalised project costs, this is contrary to the accounting principles set out in the Companies Act of 1985, as a result of which the liability associated with project finance has to be shown separately. Rolled up interest associated with the loan is capitalised in the cost pool to be depreciated once the project comes on stream.

INTERNATIONAL ACCOUNTING COMPARISONS

Introduction

Before concluding this section on oil company accounts by returning to the question of their use in the evaluation of companies, we shall look briefly at the impact of different national accounting standards on oil companies. Some of the issues concerned have already arisen in the earlier discussion of specific accounting issues.

Two categories

Users of international oil company accounts will find two categories of difference in treatment with which to cope. The first is to do with the differences between US and UK accounts on the one hand, and continental European accounts on the other, and these differences are not specifically related to the oil industry. The second is to do with the very specific differences of accounting practice which arise in the treatment of oil and gas companies, where there are some divergences between US and UK practice, and where continental European practice has generally been more lax, but is increasingly approximating to US accounting requirements.

Oil companies' disclosure

In general, the standards of accounts produced by oil and gas companies tend to be high, by comparison with their domestic counterparts in other industries. This is in large part because they are big, capital intensive companies, with a substantial need for equity. Many of the smaller capital markets, in Europe for example, are inadequate to provide them with the funding which they need. Thus, there has been a tendency for the European companies to seek listings on the London, and even more importantly the US, stock exchanges. In many cases this has coincided with partial privatisation, since many of the European oil companies were state controlled.

UK/US and Europe

With only slight oversimplification, it could be said that US and UK accounts have historically been produced for the equity investor. They attempt to represent a realistic measure of the distributable surplus which a company has generated over the year. Deferrals and provisioning, where they occur, are attempts to reflect timing differences which would otherwise distort the accounts.

Historically, continental European companies have published the accounts which they have submitted for tax purposes, which in many cases have included tax driven provisioning and accelerated depreciation. This has the effect of reducing the stated profits, which has been of little concern in the (frequent) event that the equity was mainly held by banks so as to retain some control over that portion of the company's balance sheet which concerned them more, i.e. the company's debt.

Convergence on US GAAP

With the internationalisation of equity markets, the reduction in the proportion of capital which European companies have deemed it prudent to raise in the form of debt and the more stringent requirements of US and UK investors, the distinction between Europe and the UK/US is breaking down. In some cases, companies produce accounts under two conventions: local accounts, and US GAAP (or similar). As we have already seen, Norsk Hydro is an example.

The 1991 profit and loss account of the Austrian oil company OMV (see Table 3.11) is one of the best recent examples of relative opacity. Much of the provisioning is tax driven. The treatment of inventory movements is ambiguous. Furthermore, there is no adequate split of profits between exploration and production, refining and marketing, and chemicals, despite the dramatically different factors which drive the profitability of these businesses. (The company began to provide a split by business in 1992, though the accounts remain in other ways similar to those of other Austrian and German companies.)

It remains true, however, that the major international oil companies are predominantly UK and US companies. This may be changing, as exploration and production opportunities shift from the US and the UK

Table 3.11 *OMV Group consolidated profit and loss account, year ended 31 December 1991*

	1991 *AS000*	*1990* *AS000*
1. Turnover	82 912 809	77 868 682
2. Change in stocks of finished products and work in progress, as well as services not yet invoiced	−442 187	845 287
3. Own work accounted for in fixed assets	859 167	863 966
4. Other operating income	1 378 064	1 264 389
5. Expenditure for materials and services	−44 856 405	−43 272 495
6. Staff costs	−10 424 577	−9 854 709
7. Depreciation on intangible and tangible fixed assets	−4 844 878	−4 660 239
8. Other operating expenses	−22 786 889	−20 658 363
9. Subtotal of items 1 to 8	1 795 104	2 396 518
10. Income from participating interests	97 949	126 434
11. Interest income, securities income and other income	569 498	666 123
12. Income from the write-up and disposal of financial assets	447 180	86 869
13. Expenditure arising from participating interests	−129 670	−363 058
14. Depreciation on other financial assets and on securities held as current assets	−25 425	−174 653
15. Interest payments and similar expenses	−1 037 775	−890 828
16. Subtotal of items 10 to 15	−78 243	−549 113
17. Profit on ordinary activities	1 716 861	1 847 405

Table 3.11 *Continued*

	1991 *AS000*	*1990* *AS000*
18. Extraordinary income	–	934
19. Extraordinary charges	–	−110 387
20. Extraordinary profit	–	−109 453
21. Taxes on income and profit	−292 009	−923 279
22. Profit for the year	1 424 852	814 673
23. Reversal of revenue reserves	3 148 800	1 025 226
24. Allocation to revenue reserves	−4 083 195	−1 452 516
25. Profits attributable to other companies	−10 475	−7 373
26. Profit brought forward	56	46
27. Net profit	480 038	380 056

Source: OMV.

North Sea to Russia and the Middle East, but it is essentially still the case today. Of the continental European companies, only Elf Aquitaine is of a similar size to the US majors, and Elf's accounts almost precisely conform to US GAAP requirements. As the continental European oil companies grow, and their need for capital drives them to foreign equity markets, so their levels of disclosure will tend to rise to US and UK standards.

US versus UK

US more prescriptive

Turning to the issue of the differences between US and UK accounting conventions for oil companies, it should be noted that the differences are relatively small, and in no way impede direct comparison of company performance. In general, US accounting requirements are more prescriptive than those in the UK, and the level of disclosure required is somewhat higher.

Depreciation
The greatest difference between the two conventions is with regard to depreciation of capitalised costs. While both systems use unit of production for depletion, in the US it is only proved developed reserves which are used for the computation. Estimated future development costs are not used in calculating the depletion rate. In the UK the recommended system is to use all commercial reserves, and to include the estimated costs of developing those which are currently undeveloped. Both match the costs to be depleted with the reserves against which they are depleted. The US requirement is more precise, but may exclude a large proportion of a company's commercial reserves.

Specific US requirements
Apart from the differing approaches in the UK and US to depreciation, the other main differences represent requirements which apply in the US but not the UK. Of these, the first is the requirement to produce a standardised measure of the value of future discounted cash flows, as we have already seen in the discussion of reserve disclosure.

The other two specific US requirements solely apply to full cost companies. The first is that US full cost companies are required to disclose the amortisation by unit of production for each of their cost pools. The second is that they are required to show for each cost pool the extent of the fixed assets which it contains that represent undeveloped properties, and which are therefore not being depleted as a charge to the profit and loss account. The first requirement permits an independent assessment of the profitability of each cost pool. The second prevents cost pools from being constructed in such a way that substantial unsuccessful exploration expenditure is capitalised in the same pool as valuable assets (to avoid ceiling test problems), and then not depleted as a charge to the profit and loss account.

Counter-intuitive cost pools
As we have seen, potential abuse of the choice of cost pools is one of the dangers of full cost accounting. Examples abound, but perhaps one of the most extreme was that of Tricentrol Oil and Gas, which (prior to its becoming part of Atlantic Richfield) used to operate three cost pools – one of which contained the company's interests in the 'UK and other'. Since most of the 'other', in Tricentrol's case, was represented by unsuccessful exploration efforts in the South China Seas, this choice of cost pool ensured that the exploration costs were capitalised against its UK

interests, notably the Thistle oil field in the North Sea, rather than being written off through the profit and loss account.

CONCLUSIONS

Historical cost accounts

This chapter began with the statement that there is an inevitable tension between accounts, which are concerned with reporting the past, and economic evaluation, which inevitably requires forecasts of the future. The two chapters that follow are devoted to the evaluation of exploration and production companies, and integrated companies respectively. In this chapter, a representative set of preliminary figures for an integrated company were briefly examined in Table 3.8, and those oil industry accounting conventions which have a significant impact on the consolidated main accounts – and the nature of their impact – have also been discussed.

It will have become apparent to the reader that the concept of profit as a distributable surplus generated from operations after making due provision for maintenance of the fabric of the business (the basic assumption underlying historical cost accounting) is fairly unclear, at least for the exploration and production (extraction) end of the oil industry. For refining and marketing, so long as it is replacement cost profit which is being referred to, the profit and loss account represents a better guide to performance and value.

Replacement cost unpredictable

Movement to a form of current cost accounting would not, unfortunately, help in permitting accounts to provide a useful tool for the evaluation of exploration and production businesses. While it is a reasonable assumption that the replacement cost of machinery rises with inflation, this assumption is completely inappropriate to an exploration business. It may be true that the costs of oil services rise with inflation over time, but it does not follow from this that the cost per barrel of finding oil and gas will follow this trend, for the whole industry let alone an individual company.

Use to assess track record

If the important caveat concerning the unpredictability of replacement cost is accepted, historical cost accounts, even for exploration and production companies, do record whether a company is generating revenues from its existing assets which represent a surplus over the historical capital and operating costs of producing them – whatever the timing differences which the choice of full cost or successful efforts accounting for exploration expenditure will impose.

Furthermore, historical cost accounts permit an assessment of how a company has performed over time. Any economic evaluation of assets is merely a snapshot which inevitably tends to ignore historical performance. But companies are dynamic entities whose managements can, by their decisions, enhance or destroy value. The past may not be a perfect guide to the future, but it can hardly be dismissed as irrelevant that, for example, one company has consistently found oil and gas reserves at a low cost while another has not.

CHAPTER FOUR

Valuation of exploration and production companies

INTRODUCTION

Exploration and production (E and P) companies are a good vehicle for the investor who wishes to concentrate on capital growth. This is because the yardstick by which these companies are measured is asset value. Much of the value of these companies lies in assets which will give rise to income from future production, so other conventional methods of judging performance, such as earnings multiples and PE ratios, often do not work. However, as these companies grow there is no doubt that other measures of performance become more important as distinguishing features, but the nature of the industry (which is in a constant fight against the depletion of current reserves and production) makes them an inadequate guide to future value.

Our starting point is that the foundation for all asset values must be the price that industry uses. The oil industry will always be an active trader of assets and adopts a common approach to valuation. The industry's estimate of value should therefore be the central yardstick to the equity investor. This is not due simply to the potential for outright bids for each of the companies evaluated, but also to the ability of the companies to generate internal growth in value through portfolio management (disposal and acquisition of assets).

However, the calculation of asset values to conform with industry practice is only the starting point, as the trends in both market and industry sentiment will play an important role in determining the course of share prices. The price that industry will pay for assets has in the past changed and will continue to change depending on particular circumstances. The effect of the oil price expectation is obvious but there can also be strategic considerations. For example, from 1987 to 1990 companies in the UK placed high value on assets that generated high tax liabilities in order to relieve future exploration expenditure.

Because most E and P companies are still rated against their asset value, the performance of the sector tends to be particularly volatile, and it will continue to be so because performance tends to be strongly linked to the oil price. However, at this point it would be useful to compare and contrast the performance of the E and P sector in the UK and the US over the time since the oil price crash in 1986 in order to show the impact that market sentiment can have on the performance of the sector (Fig. 4.1).

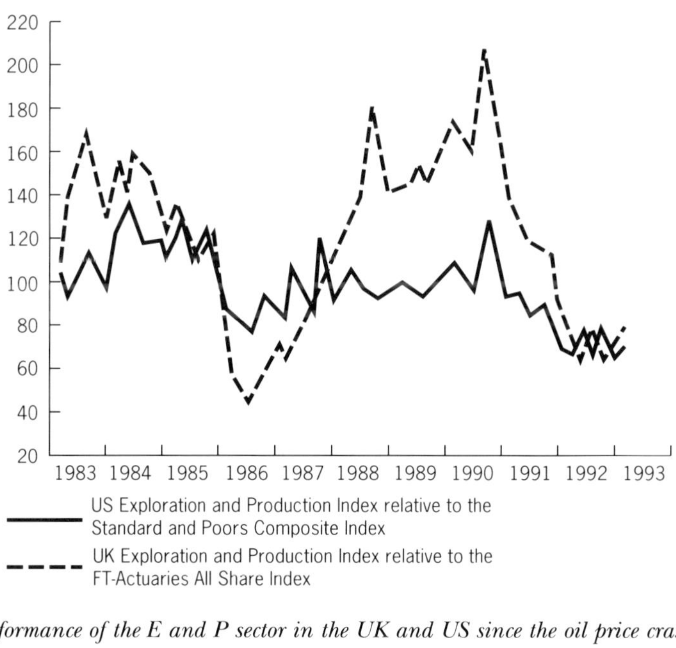

4.1 *Performance of the E and P sector in the UK and US since the oil price crash of 1986 (source: Datastream).*

Until 1986 the performance of the E and P sector in both the UK and the US tracked each other, underperforming the market as crude prices declined. When the oil price crashed in 1986 the UK sector was oversold in comparison to the US as the market took the view that the boom days of the North Sea were over. However, around the same time many oil companies began to realise that it was in fact cheaper to buy oil through take-overs and mergers than to search for oil by exploration. Therefore, from 1986 to 1990 there was a complete change in sentiment in the UK market as asset values were seen to be fully underpinned by take-overs; consequently there was a period when shares traded at very large premiums to asset values. In comparison to the US sector, the UK sector clearly became over-bought, and by 1990 the divergence had reached a point where apparent failures to realise value – combined with a massive overhang of new stock – led to a sharp fall in the value of the UK sector.

Because of the nature of E and P companies there will always be marked changes in market sentiment towards the sector. However, these changes should not be a good reason to change the basic method of valuation; after all, if the method of valuation is constantly changed there is no doubt that the use of asset values will be questioned. In fact, changes in the method of valuation may actually accentuate the volatility of the sector and give a completely misleading impression of the true worth of individual companies.

From this it is clear that the key to using asset values correctly is to maintain a consistent method of evaluation that results in values that are in line with industry. Market sentiment should only play a role in determining the appropriate premium or discount to asset value. It is ironic that although appropriate premia or discounts to asset value should only take in to account the future potential of the company evaluated, the market seems to place more emphasis on the historical track record (finding costs, management, etc).

THE INDUSTRY METHOD OF ASSET VALUATION

Asset values of E and P companies should include all of the assets that a company owns. Typically these will include producing fields, fields under development, fields under appraisal, new discoveries and exploration acreage. While differing economic assumptions will produce a range of values, the real surprises are usually only the result of major disagreements about geological interpretation, disparity in tax positions, or a strategic imperative. These are less likely to occur in the larger packages of assets where overestimations in the value of one asset tend to be cancelled out by underestimations in the value of others.

The calculation of the asset value can be conducted in a number of steps, involving a mass of data and important geological, technical and economic assumptions. The main assumptions include field-specific data such as production forecasts, capital expenditure, operating expenditure and tariffs receivable or payable. All of these assumptions are then used to construct fully taxed cash flows for each asset. It is therefore important to discuss in turn each of the input parameters, before going on to describe the method used to derive the value of the assets.

Technical and economic assumptions

Recoverable reserves

The first step in assessing the value of any company is to assess the volume of recoverable reserves present within the company's portfolio. The volume of recoverable reserves depends upon a number of factors that include:

- the total volume of the hydrocarbons in place;
- the production mechanism;
- the initial well productivity;
- number of wells;
- economic considerations, e.g. oil price.

Clearly the only time that the total recoverable reserves in an oil or gas field will be known for certain will be once the field has been abandoned and the last drop of hydrocarbons has been produced! Until that time there will be constant uncertainty surrounding the volume of reserves that will be produced, an uncertainty felt most sharply immediately after the discovery of any hydrocarbons. It is therefore little wonder that a great deal of market speculation typically follows a company announcing a discovery, as the actual size of the accumulation is rarely known for certain. Bulls of the stock may claim that a massive discovery has been made, whereas bears could discount the discovery as being too small to be of commercial value. The reality is that in the time immediately after a discovery, the range of reserves known for certain may be wide enough to encompass both points of view.

Figure 4.2 illustrates the range of uncertainties through each of the different phases of development of a discovery. It shows how the range of recoverable reserves narrows through time up until the point of abandonment. At each point in time through the development of a field it is possible to evaluate the range of possible reserves. If each of the possible ranges of reserves are arranged in order of increasing size, a special plot can be constructed with a cumulative probability scale plotted against volume. This plot is known as an 'expectation curve', and different types of expectation curve can be created to represent the various degrees of uncertainty surrounding the ultimate range in reserves volumes (see Fig. 4.3).

It is highly unlikely that the investment analyst will ever be in a position to derive an expectation curve for any particular asset. The analyst therefore relies on a company itself providing data on the volume of reserves, as well as the uncertainty surrounding that volume. The combination of reserves volume and risk gives rise to a number of reserves definitions which follow. These are the reserve definitions that are commonly used within the industry, but are different from those used for accounting purposes under *SORP 1.*

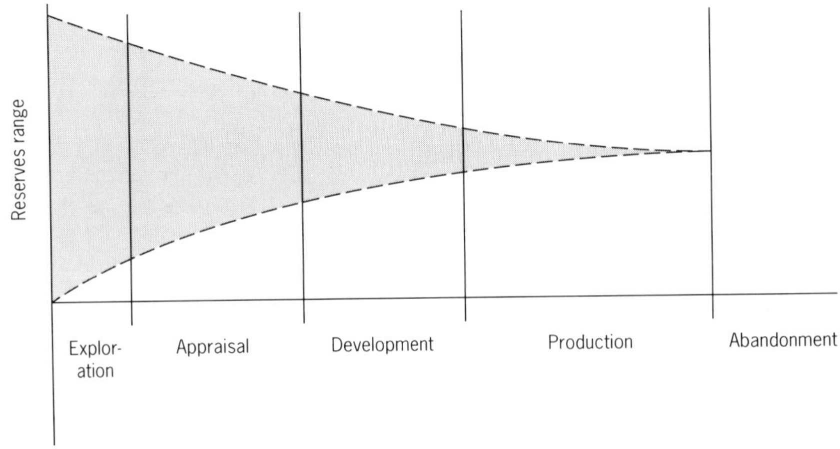

4.2 *Range of reserves uncertainty over time.*

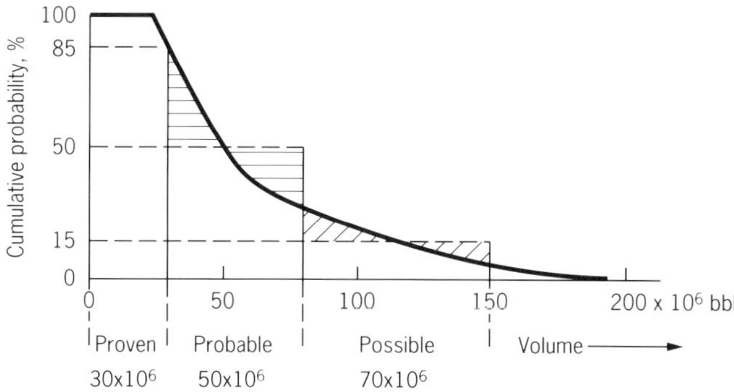

4.3 *Expectation curve used to find approximate values of proven, probable and possible reserves (source: Parker, 1984).*

- *Proven reserves* relate to the volume of hydrocarbons that geological, geophysical and engineering data indicate are in place for which there is a greater than 90% chance that the actual proven volume will be higher and only a 10% chance that it will be less.
- *Probable reserves* relate to the volume of hydrocarbons that geological, geophysical and engineering data indicate are in place for which there is a greater than 50% chance that the actual volume of proven plus probable reserves will be higher and a 50% chance that it will be less.
- *Possible reserves* relate to the volume of hydrocarbons that geological, geophysical and engineering data indicate are in place for which there is a greater than 10% chance that the actual volume of proven plus probable plus possible reserves will be higher and a 90% chance that it will be less.

The definitions most commonly quoted by industry are the proven (P90) and proven plus probable (P50) reserves levels, and it is always important to be sure exactly what number is being quoted because, as discussed earlier, there can be a big difference in volumes between the different levels. In the US most companies will only quote the proven reserves, whereas in the UK proven plus probable reserves are more often used.

The reserves definitions outlined here are based on technical consid erations alone and do not address the issue of commerciality. In the UK it is becoming increasingly popular for companies, e.g. Goal and Clyde to quote the volume of commercial reserves in their portfolios. Commercial reserves are defined as those quantities of petroleum that are considered, on the basis of information currently available and under current economic conditions, to be recoverable by present producing methods from fields which are currently in production or which have government approval for development. Commercial reserves may include probable reserves. For each E and P company it is therefore possible to build up, on an asset by asset basis, the different reserves volumes – P90, P50, etc. It is only once the range of reserves has been tabulated for each asset within a company that one can progress to the next step in the evaluation of an E and P company.

Production forecasts

The availability of a production forecast is one of the most important input parameters in the fully taxed cash flow model. However, the derivation of production profiles is an iterative process which involves information gained from appraisal wells, as well as production tests and early production history. Therefore, the confidence level in a production forecast is rather low in the early stages of evaluation, i.e. immediately after the initial exploration well, but confidence will increase through time once the field is on stream.

The reserves that can be produced during the lifetime of a field strongly depend on the production mechanism driving the hydrocarbons out of the reservoir. The most important production mechanisms – listed here – can occur either separately or in combination. Each of these mechanisms can have a profound effect on the likely production forecast that may subsequently arise.

- *Depletion drive.* This mechanism relies on the expansion of undersaturated oil or gas coming out of solution to drive the hydrocarbons out of the reservoir up to the surface. This mechanism is one of the least efficient driving mechanisms, and is often conservatively chosen as the base case for running economic evaluations.

- *Water drive.* With this mechanism water influx into the reservoir counteracts the decrease in reservoir pressure. If 100% effective, this would result in a constant rate of production for an extended period. This mechanism will occur if the water filled part of the reservoir has a large volume with good reservoir properties. Water drive can often be simulated by 'water injection', i.e. the injection of water into the reservoir below the oil water contact.

- *Gas-cap drive.* In this drive the reservoir energy is provided by the expansion of a gas-cap. This mechanism is not as efficient as water-drive and will eventually cause the wells to 'gas-out'. The drive can also be simulated by the injection of gas into the gas cap.

The production forecast obviously depends on the driving forces as illustrated in Fig. 4.4. However, another important factor is the phasing and number of producing wells that are brought on stream. If the results of a well test are not encouraging a field may need many wells, in contrast to a field that may have high productivity and which may only require one or two wells to produce all of the recoverable reserves. A production profile typically comprises three main phases, a build-up period when new producers are brought on stream, a plateau period when some new wells are brought on stream to replace the declining production from older wells, and finally a decline period when all of the producers show declining production.

In addition to the technical constraints that affect the production profile there may also be a number of external factors which although not related to the field itself can have a direct bearing on the production profile. One external constraint could be caused by portioning of production in countries that have production ceilings, e.g. current OPEC members. Another constraint that affects mainly gas fields are the gas sales contracts which often specify that the production capacity be much higher than average production in order to cope with seasonal changes in demand, or 'load factor'. In addition, gas contracts often specify that the plateau period be rather long. In new areas a major constraint might be the capacity of a trunk pipeline that connects the oil producing area with the export terminal. Such constraints in the size of the pipeline could lead to a limiting of production until the pipeline is upgraded. A specific example of this is the planned initial production from BP's Cusiana field in Colombia that is constrained by the size of the current pipeline. It is only when this pipeline is upgraded that the full potential of the field will be realised.

All of these factors have to be taken into account when considering the likely production profile of a field. Although the investment analyst is unlikely to have all of these facts available when trying to derive a production profile as discussed earlier, there are a number of sensible questions that can be asked which can lead to a reasonable forecast for any particular field. In addition, there are a couple of rules of thumb that can be applied. One rule stresses that the maximum volume of reserves that should be produced during the plateau period should not

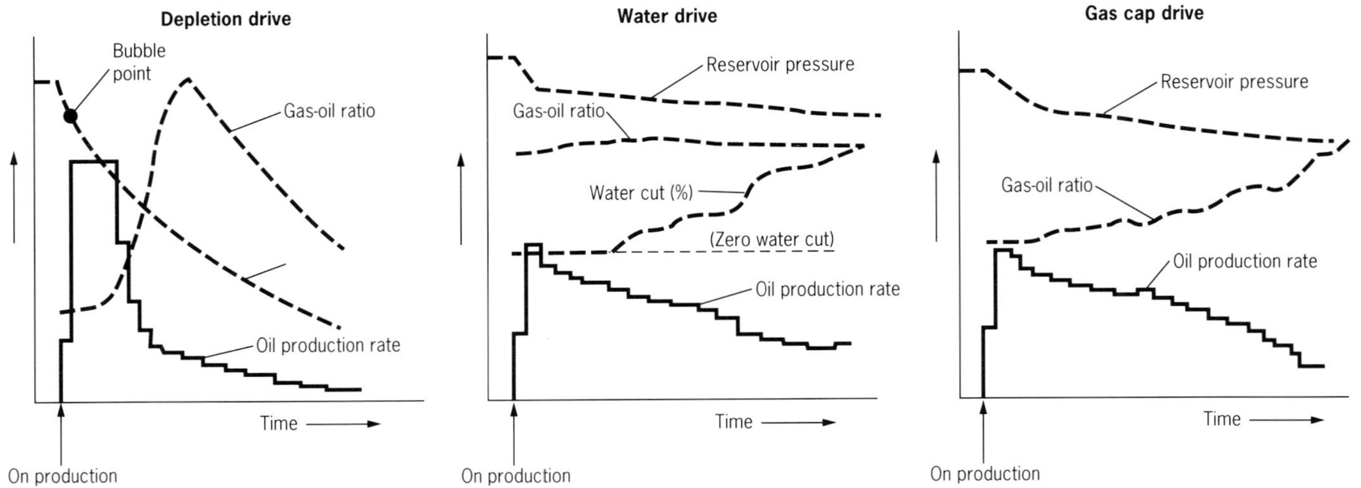

Depletion drive

Reservoir pressure: declines rapidly
Gas-oil ratio: first low, then rises to maximum and drops
Water production: negligible
Ultimate oil recovery: 5–30% (generally low)
Supplemental recovery: water or gas (re)injection

Water drive

Reservoir pressure: remains high
Gas-oil ratio: remains low
Water production: starts early and rises to large amounts
Ultimate oil recovery: 35–(75)%
Supplemental recovery: water (re)injection

Gas cap drive

Reservoir pressure: falls slowly and continuously
Gas-oil ratio: rises continuously in up structure wells
Water production: negligible
Ultimate oil recovery: 20–40%
Supplemental recovery: gas (re)injection

4.4 *Drive mechanisms and production profiles.*

exceed 12% of the total estimate of recoverable reserves. If production exceeds this amount the maximum capacities of the facilities are unlikely to be utilised for any length of time. Secondly, the rule of thumb for the rate of production decline is in the range of 15 to 20% per annum.

Capital and operating costs

With an estimate of the recoverable reserves and production forecasts it is possible to derive a development plan for any particular asset. Development scenarios and cost estimates vary during the lifetime of a project, and all costs are dependent on the timing, location and quality of the asset. The cost of any development will vary with time due to general inflation and local market conditions. The development cost will also vary between locations because of the differences in the costs of local wages, transportation from the site of construction, climatic conditions, and perhaps most importantly of all, whether the development is onshore or offshore. The quality of an asset will also have a marked impact on the cost of its development. For example, a field with high quality light oil will be much cheaper to develop than a field with heavy, sulphurous crude that will require far more facilities.

Capital expenditure, or 'capex' estimates for most development scenarios are usually based on the following criteria:

- location;
- water depth (in the case of an offshore discovery);
- terrain conditions (in the case of an onshore discovery);
- distance to the export terminal or refinery;
- recoverable reserves estimate;
- initial well production;
- reservoir drive mechanism;
- crude properties.

With all of these factors it is usually possible to use known costs from similar developments, and either scale up or scale down the costs to the size of the reserves. However, one should also take into consideration overriding market considerations. This can be illustrated by the cost of drilling, which can be very competitive and which relies strongly on the current supply/demand position. Figure 4.5 shows the relationship between oil price and drilling activity, and demonstrates that higher oil prices have in the past led to long periods of active exploration when drilling costs have been very high. After the price collapse in 1986 exploration activity was extremely depressed, and drilling costs fell quite dramatically as a result.

In addition to the overall capital expenditure required to develop a particular field, consideration must be taken of the annual operating costs. These costs differ from the development costs of the field, in that they will have to be paid throughout the lifetime of the venture. Operating costs will comprise production costs (lifting costs, treatment, workover costs, etc), maintenance costs (inspections, remedial work, etc), evacuation costs (pipeline tariffs, tanker costs, etc) as well as administrative and general overheads. All of these costs will vary from asset to asset and also from company to company. In the UK North Sea general operating costs tend to vary from around \$4/bbl (without tariffs) to nearly \$6/bbl (including tariffs). This contrasts with the operating costs in the Middle East that can be lower than \$2/bbl.

Economic assumptions

Macroeconomic factors such as oil price, inflation and exchange rates are also important ingredients in the cash flow model. Clearly each company and investment analyst will have a different view on the oil price outlook, but whatever the outlook it is important that the same

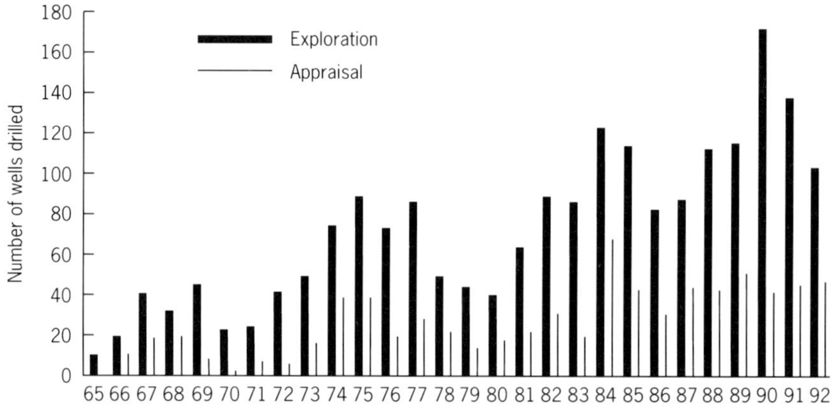

4.5 *Relationship between oil price and drilling activity, using UK offshore exploration/appraisal wells (source: Arthur Andersen).*

scenario is applied to each and every asset. It is also important to base the economic scenarios on a long term view; after all, most of the fields to be evaluated will have a life of over 10 years, so there is little point in basing the evaluation of a field on an unrealistic spike or trough in the crude oil price.

Cash flow analysis

Cash flow analysis enables an investor to see the result of an investment in a particular project; the analysis enables the investor to see how much of a cash surplus is being generated by the particular project, and therefore how much is actually available for the company to reinvest or distribute in the form of dividends. Cash flow analysis also enables the investor to see how much of a cash deficit each project is running, and so how much money has to be supplied. Every cash flow analysis comprises cash-in items (proceeds from the sale of hydrocarbons) and cash-out items (technical costs and government take).

Cash-in items

The main cash-in item is a company's share of its gross revenues, which mainly arise from the production and the sale of crude oil, natural gas or natural gas liquids (NGLs). Other cash-in items include the proceeds from services used by third parties (such as a pipeline tariff), salvage value of the assets sold after a field is abandoned (a floating production vessel, for example) or a loan from an outside financier.

Cash-out items

The initial cash-out items are the technical costs associated with the development and operation of a project. Technical costs can be divided into either capital expenditure (capex) for goods that last for more than one year, and operating expenditure (opex) for goods or services that have a lifetime of less than one year. As mentioned earlier, typical capital expenditure items are the costs of platforms, production facilities, pipelines and drilling. Capital expenditure is the dominant cash flow item in the early years of a project. Operating expenses include well repairs and workovers, lifting costs, maintenance costs and tariffs. All these costs have to be assigned to the actual period that payment is to be made.

The second – and always the largest – cash-out item is the government take (tax). The government take varies from country to country and depends on the contract terms and fiscal system operating in the host country. Generally, a company will have to pay a royalty based on the value of the recovered reserves, as well as one or more taxes based upon taxable income. Royalty is usually based upon a percentage of the gross revenues derived from the sale of hydrocarbons, and can either be paid in cash or in kind. Royalty also has to be paid from the first day of production irrespective of whether any profit has been made. However, royalty is always a tax deductible item and it is therefore fiscally treated as an operating expense.

Tax payments generally form a percentage of taxable income, and are the dominant cash-out item in the middle of the lifetime of a project. Tax is normally only paid as and when there is a fiscal profit, and it is calculated by subtracting the fiscally acceptable costs from the revenues of the sales. The definition of what are 'acceptable' costs is always laid down in the fiscal legislation of the host country. Royalties and operating expenses are normally included in the allowable fiscal costs in the year in which the expenses have been incurred.

An additional item among the fiscal costs which can usually be deducted is depreciation. Depreciation is calculated according to the rules set by the tax legislation of the host country, and these are generally divisible among the three following categories:

- *Straight line method.* The assets are depreciated in equal amounts over a number of years, e.g. 20% of the value of the asset annually over 5 years.
- *Declining balance method.* Each year a fixed percentage, e.g. 20%, of the non-depreciated net book value as at the beginning of the year is to be depreciated. This method results in having non-depreciated assets at the end of the lifetime of a venture.
- *Depletion method.* With this method, the depreciation is equal to the ratio of the annual production to the commercially proven reserves at the beginning of the year, multiplied by the non-depreciated assets at the beginning of the year.

Since depreciation is one of the factors that helps determine the taxable income, a higher level of early depreciation will lead to a lower tax bill in the early years of a project, thereby improving the project's early cash flow and overall economics. Of the three methods of depreciation outlined, the straight line method generally leads to the best results.

Net cash flow

The net cash flow represents the residue once the cash-out items have been subtracted from the cash-in items. In general, the net cash flow can be defined as follows:

$$\text{Net cash flow} = \text{Revenues} - \text{Capex} - \text{Opex} - \text{Royalty} - \text{Tax}$$

This compares with the profits that are defined as:

$$\text{Profit} = \text{Revenues} - \text{Depreciation} - \text{Opex} - \text{Royalty} - \text{Tax}$$

The relationship between the net cash flow and the profits are shown in Fig. 4.6, and from this it is easy to observe that if the capex exceeds the depreciation, the net cash flow will be less than the profit after tax. Alternatively, if the depreciation exceeds the capex, the cash flow will be larger than the profit. At this point it is worth stressing that profit and depreciation are only accounting terms and are not cash flow items.

Cash flows are usually represented as bar charts from which only a few economic indicators can be derived (see Fig. 4.7). On its own it only serves as a planning document, informing the investor of exactly how much he is expected to contribute to any particular project, as well as how much he can expect to receive in any particular year. On its own, the net cash flow graph will only provide an indication of when a particular project is likely to turn cash positive, and provide information on the length of time that the project is likely to last before abandonment.

More information can be derived from the cumulative cash flow, which is defined as being the running sum of the net cash flow (see Fig. 4.8). From the cumulative cash flow graphs, it is possible to determine three main economic parameters:

- *Maximum exposure:* the maximum value of the cumulative cash outflow. This value helps the investor assess the maximum amount of money at risk and is of particular importance to investors in small E and P companies (such as Aran in the UK in 1992, which was exposed to two very large development projects that would take gearing from 0% to over 100% in less than one year prior to any cash receipts).

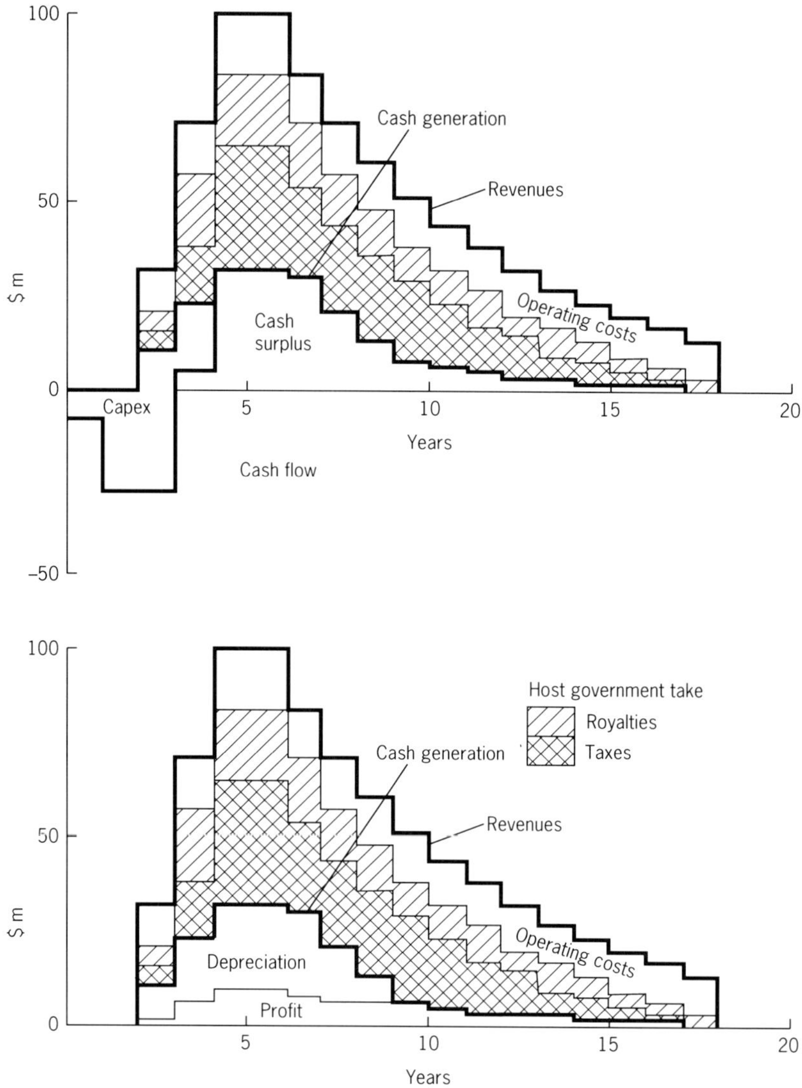

4.6 *Breakdown of revenues of a medium cost oil field, showing the relationship between net cash flow and profit, from the view of an economist (a) and an accountant (b).*

- *Pay-out time:* the time at which point the cumulative cash flow turns positive and the investor recovers his original investment.
- *Ultimate cash surplus:* the value of the cumulative cash flow at the end of the life of a project.

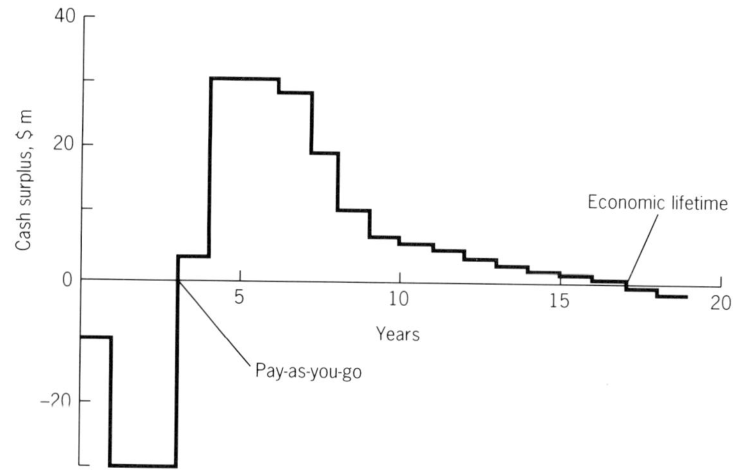

4.7 *Typical annual cash flow, shown as a bar chart (source:* Van Dam, 1980).

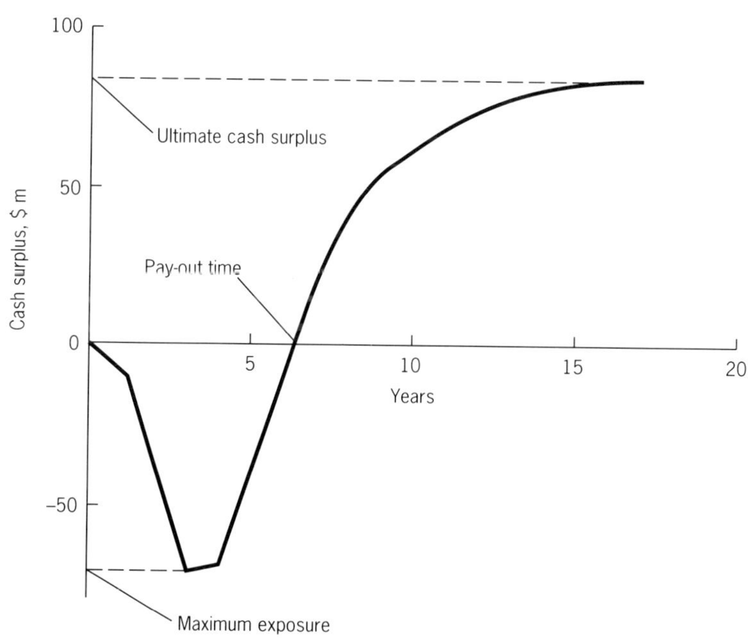

4.8 *Example of cumulative cash flow shown in graph form (source: Van Dam, 1980).*

Discounted cash flow analysis

Net present value

Most of the important economic indicators are derived by discounting the net cash flows, and the process by which the profitability of a project can be determined is known as 'discounting'. Using this method the analyst will be able to calculate the net present value (NPV) of any project. In general, companies should accept the projects that maximise their NPV, and reject those projects which have a negative NPV.

Discounting the cash flow is useful because it helps the investor choose between alternative projects. For example, money could be invested in a bank account or in government bonds and earn a guaranteed rate of return on the initial investment. In order to be convinced of investing in a particular project the investor has to be convinced that it will yield a higher rate of return than the best alternative low risk investment.

In addition to the base case, oil and gas projects are always exposed to a variety of risks that might lead the investor to have doubts about whether the promise of reward later will actually come true. With E and P companies there is the initial exploration risk, followed by the risk that the reserves are much smaller than originally anticipated. There may also be technical risk (higher than anticipated construction costs), economic risk (lower than forecast oil prices) and political risk (changes in taxation or nationalisation). All of these risks have to be considered when evaluating the additional profits that should be made from a project in comparison with the best alternative low risk investment opportunity.

The present value of any amount to be received at a specified future date is equivalent to the amount that would have to be put on compound interest at the reference date to yield the required amount at the future date. The factor by which an amount has to be multiplied to arrive at the present value is known as the 'discount factor', and the annual rate is known as the 'discount rate'. It is obvious that disclosing a present value without the reference date or the discount rate is meaningless. The net present value of a project is the sum of the present values of the cash surpluses or cash deficits over its lifetime. It represents the value to the investor who will also earn interest in the project at an amount greater or equal to the discount rate used, providing the sum of the present values is positive.

The NPV can therefore be taken quite literally, as it represents to the investor the value of 'owning' an existing project. This is important, because the owner should be willing to sell the project provided the price is higher than his own NPV. Similarly, a potential bidder should never pay more than the NPV in order to acquire a project. This concept is vital when discussing the appropriate premium and discount to asset value on which company shares should trade on.

The present value cash surplus/deficit of a cash flow can be plotted against the discount rate in order to obtain the 'present value profile' (see Fig. 4.9). For lower discount rates the present value is usually positive and for higher discounts it usually turns negative. The discount rate for which the present value equals the zero is known as the rate of return of the project. For as long as the discount rate of the investor is less than the expected rate of return the project will be economically viable. This point is important, because if the NPV of a project is equal to zero, it does not mean that a project is worthless – merely that it is yielding a rate of return equal to the discount rate used. If the NPV is positive, the sum represents the present value of the cash worth in excess of making a return equal to the discount rate. Put another way, a positive NPV is the amount of additional money that could be invested at time zero in the project, and still realise a rate of return equal to the discount rate used for the total investment.

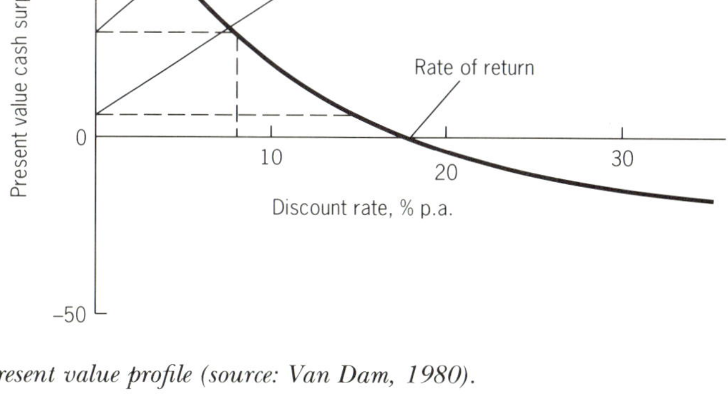

4.9 *A present value profile (source: Van Dam, 1980).*

applies to exploration acreage; this subject is specifically dealt with from page 143. The value of exploration acreage can only be proven following the results of future drilling.

Consolidation of cash flows
It should now be clear that the NPVs derived for individual assets each comprise a fully taxed cash flow. At this stage, however, it is very important to bear in mind the fiscal position of the company; if a company already has large fiscal income and is consequently paying tax, it is often possible to expense costs and depreciation of new projects against the existing income from other projects. In this way a reduction in overall annual tax payments can be achieved, and these can be regarded as part of a company's overall cash flow in the form of negative tax payments in the relevant years. In contrast, if a new project is a company's first or only one, depreciation can only commence at production start-up, and eventual fiscal losses have to be carried forward to later years when they can be absorbed against future profits.

In general there are two types of tax credit that can be added to an E and P company's asset value:

1 *Tax credits arising from historical expenditure:* These credits arise from expenditure that has already occurred and will benefit the company even if no further expenditure is incurred. Although it may be difficult to assign them to particular assets it is possible to calculate the overall value of the tax synergy that results from the consolidation of all of the assets. In the UK the impact of historical tax credits can be quite large because of the effects of historical allowable expenditure.
2 *Tax relief for future exploration expenditure:* Many companies can relieve a substantial portion of their exploration expenditure with current tax liabilities. In fact until April 1993 a fully tax-paying company in the UK could relieve up to 83% of the total cost of any exploration expenditure. It is therefore important to incorporate into the asset value any tax synergy that may arise from offsetting future exploration expenditure.

Tax regimes

Since tax is often the largest cash-out item in the cash flow calculation it is worth discussing the different type of tax systems that exist worldwide.

The basic example used earlier is commonly known as the 'royalty/tax' system. In practice, however, there are currently four main types of contract in use.

The royalty/tax system is the oldest type of contract and could be described as a 'joint venture contract without participation'. In the 1960s, national oil companies were set up to participate in oil ventures as a vehicle for governments to get some influence in the development of their national resources. Originally, the level of state participation was rather low, but in 1974 the OPEC countries increased participation to 60% . In joint venture contracts, a private oil company owned the oil in the ground that it had discovered. This aspect of the contract was found to be unacceptable in many countries, and for this reason two new types of contract emerged: 'production sharing' and 'risk contracts', whereby the private oil company is put into the role of a contractor who invests in exploration, appraisal and development, produces the oil for the national oil company, and receives a reward, either in cash or in kind. In 'technical service agreements', the private oil company will undertake research studies and offer training for which it will receive cash. In this agreement, the private oil company is not investing heavily; it is merely providing a service to the national oil company.

The choice of a particular contract type is related to the oil supply balance and the economic strength of the country involved. Typically, production sharing contracts are applied under all types of conditions. Joint venture contracts are given as much application in weak oil exporting countries as in strong oil importing countries. Risk contracts are usually only applied in economically weak oil importing countries, and technical service agreements only in strong oil exporting countries.

The different tax agreements are all discussed separately in the following sections. It should be stressed, however, that in order to stimulate the development of marginal fields the effective rate of tax may change. Similarly, in the current environment of relatively low oil prices many governments often make an effort to see if certain special tax exemptions may be applied if developments that are economically attractive pre-tax become loss-making on a post-tax basis.

Joint venture contracts

With joint venture contracts a national oil company is entitled to participate in the development and exploitation of the fields in a specified acreage. There can be some difference in the level of participation as well as in the pre-development costs that will be incurred. In Norway the production potential of the oil or gas field to be developed has been taken into consideration with the private industry share being only 20% once production levels reach 400 000 bbl/d. In contrast, in the Netherlands the national oil company backs in for 50% of the venture. In the Netherlands exploration costs are reimbursed (without interest and in the case of success only) whereas in Norway exploration costs have not been shared.

In some cases the national oil company may not have any funds available to fund the development phase, and in such cases it may have to be 'carried' through the development phase. This means that the private oil company will have to pay the national oil company's share of the development costs, which should then be repaid out of its share of revenues.

In joint ventures, taxation is based on the royalty/tax system. Royalty payments are usually based on the revenues. Frequently, more than one tax has to be paid, and in these cases the various taxes will each have different definitions of fiscal costs that are allowable, as well as different depreciation rules. In some countries taxes may also be levied on a tax reference price or posted price, which may actually be higher than the price actually received for the oil. This is a particularly harsh way of increasing taxes, as there can be no corresponding increase in the tax relief available from operating costs and depreciation.

In some other countries a 'windfall profits tax' has been introduced. The purpose of this tax has been to tax away any extra profits that result from a faster than normal increase in oil prices. In this case the private oil company would pay a 'normal' tax on the profits that had increased (or decreased) proportionally to a suitably chosen index, e.g. the consumer price index. The difference between these theoretical profits and the actual profits are regarded as windfall profits and are subject to an extra high tax rate. With the relatively low oil prices of recent years, windfall profits have generally become a less common phenomenon.

Production sharing contracts

With production sharing contracts all rights to the actual petroleum resources in the ground are vested to the state. Authority is then granted to the national oil company to conclude an agreement with the oil industry, which is then considered to be a contractor. The agreement is characterised by the following provisions:

- The contractor works under the supervision of the national oil company.
- The contractor has to provide all the funds for the operations and carries all the risks involved.
- All of the assets will accrue to the national oil company, either immediately upon acquisition, or gradually with depreciation.
- When the venture reaches the producing stage the contractor may recover his costs out of a share of the production up to a specific maximum. This share is called 'cost oil'. The manner in which the costs can be recovered within this maximum, and which costs qualify for such recovery, will be detailed in the agreement.
- The balance of the available production – called 'profit oil', which is what remains of the production after the contractor has recovered its costs – is shared or split between the national oil company and the contractor in agreed proportions; these proportions sometimes depend on the production levels achieved.

There was a time when all taxes and royalties were paid by the national oil company on behalf of the contractor; the latter would receive its profit oil free from any fiscal burden. However, it is now more common for the profit share to be 'grossed up' for corporation tax, in which case the private oil company's profit oil is subject to that tax. This construction usually means that double taxation can be avoided.

In addition to these taxes, the contractor might also be required to supply, on a pro rata basis, the local domestic market with a small quantity of hydrocarbons. This 'domestic supply obligation' reduces the amount of crude that can be exported by the private oil company, and may have economic consequences if, as is usual, the domestic market is to be supplied at a vastly reduced price. A simple procedure to calculate a production sharing cash flow is outlined in Table 4.1.

In the presence of inflation, production sharing contracts with a low cost oil maximum combined with high operating costs may result in a situation in which the capex might not be recoverable in cost oil. In such cases the profitability indicators can be very sensitive to operating costs.

Risk contracts

With risk contracts all rights to the petroleum resources are vested in the state, while the private oil company acts in the role of a contractor,

Table 4.1 *Procedure to calculate production sharing contracts, $m*

Year	*1*	*2*	*3*	*4*	*5*	*6*	*7*	*8*	*9*	*10*	*11*	*12*
1 Capex	50.0	50.0										
2 Revenues	0	0	99.7	89.7	80.8	72.7	65.4	58.9	53.0	47.7	42.9	38.6
3 Fixed Opex	0	0	5.0	5.0	5.0	5.0	5.0	5.0	5.0	5.0	5.0	5.0
4 Variable	0	0	5.0	4.5	4.0	3.6	3.3	2.9	2.6	2.4	2.1	1.9
5 Total	0	0	10.0	9.5	9.0	8.6	8.3	7.9	7.6	7.4	7.1	6.9
6 Cost oil max (COM)	0	0	34.9	31.4	28.3	23.4	22.9	20.6	18.5	16.7	15.0	13.5
7 COM-Opex	0	0	24.9	21.9	19.3	16.8	14.6	12.7	10.9	9.3	7.9	6.6
8 Capex recovery	0	0	24.9	21.9	19.3	16.8	14.6	2.5	0	0	0	0
9 Non-recoverable capex	50.0	100.0	75.1	53.2	33.9	17.1	2.5	0	0	0	0	0
10 Used cost oil	0	0	35.9	31.4	28.3	25.4	22.9	10.4	7.6	7.4	7.1	6.9
11 Profit oil total	0	0	64.8	58.3	52.5	47.2	42.5	48.5	45.4	40.3	35.8	31.7
12 POC share	0	0	9.7	8.7	7.9	7.1	6.3	7.3	6.8	6.0	5.4	4.8
13 Cash-in	0	0	44.6	40.2	36.1	32.5	29.3	17.7	14.4	13.4	12.5	11.7
14 Cash-out	50.0	50.0	10.0	9.5	9.0	8.6	8.3	7.9	7.6	7.4	7.1	6.9
15 Cash surplus	−50.0	−50.0	34.6	30.7	27.1	23.9	21.0	9.8	6.8	6.0	5.4	4.8
16 Cumulative cash surpluses	−50.0	−100.0	−65.4	−34.7	−7.6	16.3	37.3	47.1	53.9	59.9	65.3	70.1

working under the supervision of the national oil company. All oil produced will have to be delivered to the national oil company at an agreed price, which is normally much lower than the international oil price.

The private oil company will be subject to the local taxation but no special petroleum taxes will be levied. Operating costs can usually be expensed and capitalised assets have to be depreciated according to the local rules.

Comparisons

The rate of return on a joint venture is often higher than that for production sharing agreements and risk contracts. The United Kingdom Continental Shelf (UKCS) and Norwegian tax regimes are certainly more attractive than the production sharing contracts of Indonesia. However, any comparison is very difficult because most countries will try to introduce measures to stimulate the development of their natural resources, particularly those resources which may only be marginally economic. One such measure was the introduction of 'oil allowances' in the UKCS; as a result of this many small fields were exempt from paying petroleum revenue tax (PRT).

In contrast, many countries are very sensitive to the charge that foreign oil companies are profiteering and have thought it necessary to introduce an 'additional profits tax'. With such contracts, either the tax rate or the national oil company's share of the profit oil is increased as soon as the rate of return on a project exceeds a certain level. This procedure guarantees a decent rate of return for the investing private oil company, but it also prevents excessive profits. This form of tax is present in countries such as Papua New Guinea and Tanzania.

EXPLORATION ACREAGE

Exploration acreage is the most difficult asset within an E and P company's portfolio for an analyst to value, yet it is perhaps the most important since it is mainly through successful exploration that substantial value can be added. However, without any detailed information on individual exploration prospects any valuation will be very subjective, depending very much on the level of detail that individual companies supply.

Valuation of exploration acreage should take into consideration the geological prospectivity of the acreage, the size and number of prospects that remain to be discovered, the associated risk, and the potential economic value. This analysis is always undertaken by oil companies when valuing acreage for farm-ins and farm-outs, deciding where to drill and looking at acquisitions and disposals. By their very nature the results are not expected to be accurate, but they do provide industry with an internal guide to the value of acreage, which in turn creates market value.

For the analyst the valuation of exploration acreage is very subjective, and often depends on a knowledge of the historical levels of activity within a certain area. For example, it may be possible to value prospective acreage in the North Sea on a block-by-block basis because of the intensive exploration programmes that have already taken place, and because of information on the likely size and location of prospects that remain to be drilled. In more remote areas, where exploration programmes have not even commenced, it is very difficult to value acreage without any knowledge of the possible size of prospects or the geological risks involved. Above all, the evaluation of exploration acreage does involve some understanding of the link between risk and reward using the concept of 'expected value'.

Concept of expected value

The economic evaluation of an undrilled exploration prospect can be simplified by trying to estimate the expected monetary value (EMV). The EMV represents the value of an uncertain opportunity in terms of money, and allows comparison between different opportunities on the same scale. The EMV also allows account to be taken of the perceived risks in different prospects.

Prior to the drilling of an exploration well the value of the prospect is usually very uncertain because of the range of possible reserves that may be discovered, combined with the economic and technical uncertainties. However, although it may be impossible to predict the actual volume of reserves that exist in the ground it may be possible to estimate the ranges of possible values that each variable may have, even though it will be impossible to identify the one outcome that will actually occur. There are four main ways for treating risk and uncertainty in the evaluation of exploration prospects:

1 *Sensitivity analysis:* This method evaluates the new profitability for any percentage change in a given parameter. However, it does not provide any information about the likelihood that this percentage change will actually occur.

2 *Adjectives:* This method enables companies to communicate information about the risk to analysts, e.g. high risk, low risk, etc. However, the information is subjective and what may be regarded as low risk to one is high risk to another, and will always reflect personal or corporate bias.

3 *Higher discount rates:* Discount rates are not an acceptable way of treating risk and should not reflect geological uncertainty. Risk should be accounted for by explicit probability numbers, and discount rates should be exclusive of technical risk and uncertainty (only incorporating financial risk and alternative business opportunities). There is a real danger of doubly penalising an exploration prospect by first applying geological risk and subsequently using a higher discount rate to cover the investment in such a 'risky' project.

4 *Risk analysis:* Risk analysis is the quantification of risk with probabilities. Measurement of the relative likelihood of occurrence of each possible outcome can be made on a 'zero to one' scale (zero means the event cannot occur, 0.5 means the event has a 50% chance of success, etc). The NPV of each possible outcome can then be weighted against its likelihood of occurrence. This is the expected value concept and from the weighted average profitability the EMV – which takes into account both success and failure – can be determined.

The EMV can be defined as the expected net present value of success less the expected cost of failure.

$$\text{EMV} = \text{Probability of success} \times (\text{Unrisked NPV} - \text{Exploration costs}) - (1 - \text{POS}) \times \text{Exploration cost}$$

This can be further simplified to produce the following formula:

$$\text{EMV} = \text{POS} \times (\text{Unrisked NPV}) - 100\% \text{ Exploration costs}$$

In order to define the EMV three different types of information are required:

1 Possible outcomes that could occur if the well is drilled, e.g. dry hole, reserves levels, etc.

2 Probabilities of success associated within each possible outcome.

3 NPV for each possible outcome.

Using this information the weighted average of all the outcomes will lead to the EMV and clearly different prospects will have different outcomes as defined by their own specific expectation curves (Fig. 4.10). In practice it may only be possible to take a few – three, say – possible outcomes which are representative of the total range of possible outcomes. However from an economic point of view, once an EMV has been calculated using the formula given, the decision rule is rather simple:

- If the EMV > 0 proceed and add value
- If the EMV < 0 reject or lose value

With a whole portfolio of exploration prospects, companies should select the alternatives which maximise the EMV. The average EMV of a

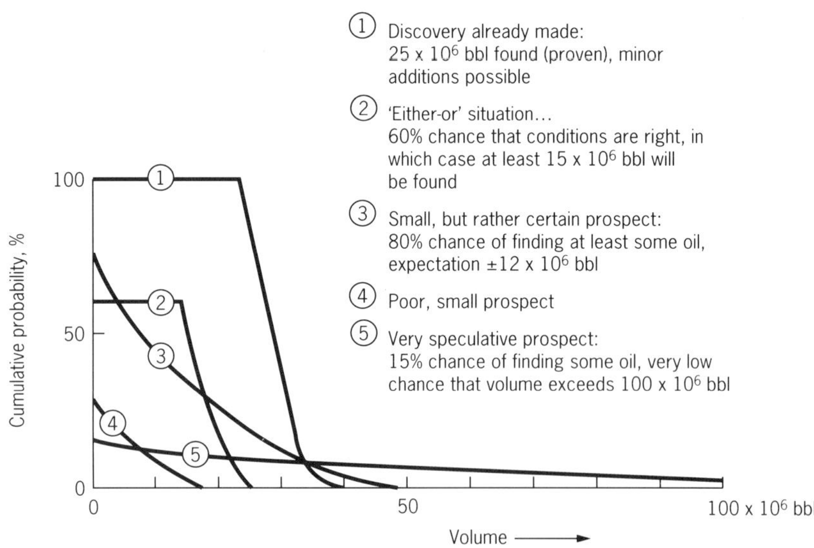

4.10 *Expectation curve (source: Parker, 1984).*

portfolio of exploration prospects is a probability weighted NPV and it represents the average gain per decision that a company would realise over a series of repeated trials; the notion of repeated trials is essential to this definition. For the EMV to be equal to the average gain per decision, it is essential that the trial is repeated in order to balance the losses against the gains. Typically, the number of trials must exceed twenty for this approach to be meaningful. It is also important to note that the decisions do not have to be fully company owned, nor do they have to be company operated wells.

In exploration it is not possible to 'redrill' a single prospect several times in the hope of finding a different outcome. However, the repeated trial concept can be overcome if each prospect is treated as one of many corporate decisions. Therefore if a company participates in many drilling decisions and consistently maximises the positive EMV from among the choice of prospects to be drilled, it follows that the value to be added from the results of a number of exploration wells should converge to the sum of all of the EMVs.

The main advantage of using the EMV concept is that it takes account of both geologic and economic risks and enables high risk and low risk prospects to be evaluated on a common basis. The method also considers a whole range of possible outcomes and means that there is no need to make further sensitivity analyses. The method also allows geological and economic studies to be cast into a form which can be used as a basis for the valuation of a company's portfolio of acreage.

The biggest problem with the application of the EMV concept is that although in the long term the EMV number has a quantitative meaning, it means very little on its own for one particular prospect. In addition, the concept also demands that a number of drilling decisions are undertaken, and this may not be possible for companies with small or limited drilling budgets. There will also be difficulties in quantifying the exploration and long term economic risks, and for many companies there may also be strategic issues which may obstruct choosing the prospects with the highest EMVs.

Acreage valuation

Prospect-by-prospect analysis

As an analyst it is very difficult, if not impossible, to obtain all of the information required to carry out a full and thorough valuation of exploration acreage based on a prospect-by-prospect analysis. Companies are often prepared to divulge the details of the prospects that they intend to drill in any particular year, but they will certainly not give details of the total prospect inventory. They might also give details of the mean success volume that they anticipate in the event of success, with a subjective opinion about the probability of success. Although this might not be enough to give a full and valid valuation of the acreage, it is possible to use this data within a framework of parameters that can be determined by an analyst. Such a framework might include:

- The minimum economic volumes required.
- The steepness of the NPV versus reserves graph.
- The minimum probability of success required to give a positive EMV.

It is simple to derive the minimum economic volume required in a particular basin or sub-basin; depending on the level of knowledge about a particular basin it may even be possible to determine various levels of minimum economic volume depending on the type of play or location. For example, in the North Sea the minimum volume required to support a full standalone development can be over 100 mmbbl, whereas a satellite development tied into existing facilities might only require 25 mmbbl, and a single well tie-in might only require 8 mmbbl. In contrast, the minimum volume required to support a new onshore development in the US is likely to be very small (1–2 mmbbl). With a framework of minimum economic volumes it is possible to put into context the size of the prospects that a company will be drilling. If the mean success volume of the prospect to be drilled is less than the economic threshold, then it is unlikely that the prospect will add value, while the prospect is larger than the minimum economic threshold, then clearly the prospect could add value.

The unrisked value of the prospect to be drilled depends largely on the steepness of the NPV versus reserves curve, which in turn will depend on a number of factors, e.g. the local tax regime. The steeper the curve, the greater the incremental value for reserves over and above the minimum economic volume – provided the exploration well is successful.

It is also possible for the analyst to determine the minimum probability of success that is required to generate a positive EMV. The minimum probability of success is a useful number as it puts into context the likelihood of a company achieving its objective. After all, if the minimum probability of success required to discover a billion barrel field is calculated to be 50% and the company considers the prospect to have only a 10% chance of success, the EMV of the prospect will be negative.

A limitation for the analyst in the evaluation of exploration acreage using prospect-by-prospect analysis is the fact that the EMV only represents the additional value that a company will achieve if all of its prospects are tested. With companies generally only prepared to divulge a one year drilling programme it is very rare that a whole portfolio of prospects can be evaluated. Therefore, for most of the larger E and P companies it is impossible to value the whole portfolio of exploration prospects unless a full team of explorationists is independently employed to interpret each company's acreage, a luxury that no analyst can possibly afford. In contrast, the EMV of a smaller company's acreage can be statistically meaningless if the number of prospects within its portfolio is quite small (less than twenty).

Despite all of the limitations, it is possible to determine an EMV for specific prospects (following the method outlined previously), and in calculating this value it is possible to determine the added value that may be achieved in the event of success for a number of drilling decisions. The EMV method is also the best method for evaluating the exploration potential of any company, and should be adopted wherever possible.

Basin-by-basin analysis

In order to evaluate acreage on a basin-by-basin approach it is necessary to obtain a historical record of the results of exploration. This record can be plotted using a creaming curve which shows that as exploration proceeds the largest fields will be discovered first, and that with time an increasing number of wells will be required to discover the remaining reserves in smaller and smaller fields. In the UKCS it is possible to draw these creaming curves for different parts of the North Sea (see Fig. 4.11); the results of this show that in the northern and southern North Sea only relatively small prospects remain to be discovered. However, the central North Sea still lies on the steeper part of the creaming curve, and there are still plenty of large prospects to be discovered. In relatively unexplored overseas basins which are known to contain untapped hydrocarbons, early exploration gives many companies the opportunity to discover them in large volumes. However, in such areas it is very difficult to predict the ultimate volume that might be produced from each basin, and analogies will have to be drawn with mature geological provinces of a similar size and setting.

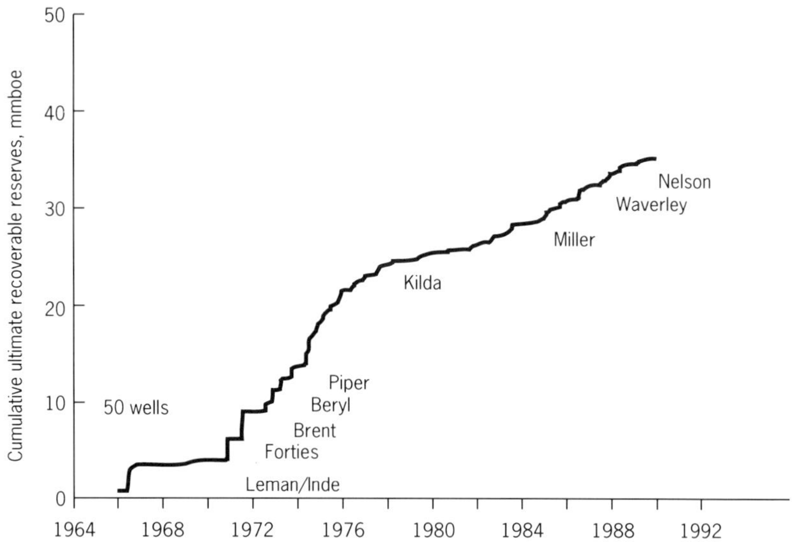

4.11 *Creaming curve for UK North Sea hydrocarbon discoveries.*

Using creaming curves it is possible to estimate the volume of reserves that remain to be discovered in a particular basin and by dividing a company's net acreage position by the gross acreage in the basin, it is possible to obtain a very crude value of the volume of potential reserves in which the company could be expected to share. Using NPV versus reserves curves calculated from developed reserves in the basin, it is possible to assign a value per barrel to the potential reserves that remain to be discovered, and then to crudely estimate the company's exposure to the basin based on its net to gross acreage position.

Value per acre

The valuation of acreage using the value per acre method is still the most common method that analysts use to assign value to exploration acreage; it can be based either on the market value of the acreage if it were to be traded, or on the value of commercial reserves per acre discovered so far.

Acreage valuations based on the market value of acreage is very common in the US, where a system of lease bidding operates in licence rounds. Economic evaluations by companies of the prospects within the blocks for which they intend to bid leads to a valuation of the block.

Since the value of the successful bid is public knowledge it is possible to gain an idea of the average value per acre that companies are prepared to pay for acreage.

Market value per acre can also be estimated in most other parts of the world, based on the level of commitment that a company is prepared to bid in order to explore. In the UKCS where there are licence rounds every two years, the licences are awarded on the basis of the level of exploration activity that companies are prepared to undertake. Typically this can consist of one or two exploration wells plus a seismic survey. This level of commitment can be costed, and it must be assumed that the company believes the risked NPV of the block is higher than this cost; otherwise, it would have been foolish to bid in the first place. It is therefore possible to build up an inventory of the average level of commitment and translate this into a value per acre figure. Similar exercises can be carried out on acreage in other parts of the world.

The problem with using the market value per acre methodology is that it does not take into account – or even attempt to assess – the actual geological prospectivity of the acreage, nor does it address the real value that can be added if a number of prospects drilled are successful. It merely assumes that if the company trades its acreage this is the average market value that might be received. Furthermore, the market value method will also fail in new frontier areas where typically vast areas of acreage are often awarded which cannot possibly be fully exploited within the terms of the licence, and which in any case might only be prospective in a small area. In such cases it would be highly misleading to use market values per acre as it would result in unprospective areas being assigned value and prospective areas being completely undervalued.

An evaluation of acreage based on the commercial value per acre that has previously been discovered in a basin is a better way of valuing acreage, as it does at least attempt to quantify the actual drilling results and turn them round into an actual prediction of the value that might be added, on a statistical basis, if a company were to explore that block. The value per acre is derived by looking at the net present value of all of the discoveries within a given area, e.g. the northern North Sea, and then dividing this value by the area in which the discoveries were made. The limitation of this method is that it assumes that success rates will continue into the future, whereas in reality it is quite likely that success rates will fall and the discoveries become smaller and less valuable. Despite this, the method does at least attempt to look at the value that can be added to a company, and in averaging out the value over an area which contains both discoveries and dry holes it attempts to take into account the dry hole risk. In new ventures where there is no previous drilling history it will be impossible to apply this method.

Historical record

A more realistic way (for an investment analyst) of looking at the upside potential and value of the exploration potential of an E and P company is to combine historic finding costs with future expenditure plans, in order to assess the potential volume of reserves that might be added to a company within a given period of time. An estimate of the value of these reserves can then be made.

For each E and P company it is often possible to derive a broad record of finding costs over time. The finding cost is simply the volume of additional reserves found through exploration, extensions and revisions each year, divided by the total exploration cost. On an annual basis the figure can be quite meaningless; however, a five year running average will begin to give a fair reflection of the track record of a company's ability to explore and discover new reserves.

If the future exploration budget of a company is known, it is possible to combine it with the historic finding cost in order to derive a figure for the volume of reserves that might be expected to be added through time. Although it is difficult to know exactly how long to project the exploration expenditure into the future, it seems reasonable to adopt a five year time horizon, that being the average time that it takes for an E and P company to turn over its acreage portfolio. However, the limitations of this method are obvious. Forward projections of historic finding costs assume that a company will continue to discover with the same rate of success, which will be difficult if it is running out of prospective acreage.

The next step is to translate the volume of reserves into a value. This will have to be done using an average value per barrel, as it is impossible to predict exactly where the reserves will be discovered – unless the company is prepared to divulge the split in the spend on a country-by-country or block-by-block basis.

Once the value of the reserves has been determined the full cost of the exploration programme required to discover the reserves needs to be subtracted from the total. Obviously, the value of acreage which belongs to a company with a poor track record of exploration will be much lower than the value of acreage which belongs to a good explorer (if this method of valuation is the only one to be applied).

Conclusions

Exploration acreage is the lifeblood of the E and P companies from which significant additional value can be added. However, as has been explained already, there is no simple way for an analyst to have access to the detail required to value acreage without the full co-operation of the company which owns it – this is rare indeed. The best way to value acreage is on a prospect-by-prospect basis, but in the absence of any detailed information the most effective way to evaluate the potential of the exploration effort must be to take a view on the additional reserves that a company can add through time, based on its historic finding costs. Although this is less than satisfactory, for the reasons outlined earlier, it does give a more tangible feel to the additional value that can be added. It also allows sensitivities to be run in order to examine the impact of lower or higher finding costs.

ASSET VALUE SENSITIVITIES

Having derived the value of all of the assets in a company's portfolio, it is also necessary to run sensitivities on all of the values. Sensitivities in oil price scenarios can obviously have a big impact on asset values, although the degree of changes can be diluted by the effects of taxation. Typically, an increase in the oil price from around $17/bbl up to $20/bbl can lead to an increase in value of about 24%. In contrast, a fall in oil price assumptions usually leads to a much larger fall in value. The reason for this difference is that with an increase in the oil price there will always be an increasing tax burden, while a fall in oil prices may actually lead to smaller discoveries becoming uneconomic. The impact of oil price sensitivities are important, because short term changes in the oil price always lead to changes in the value of shares, and the analyst must be aware of the impact of sustained changes in the oil price on asset values. However, the analyst should resist the temptation to continually alter asset values on the basis of short term movements in the oil price, as this will lead to the accusation that 'asset values only follow share prices'.

ASSET VALUE COMPONENTS

The main limitation of using asset values in isolation is that they only give a snapshot of a company at a given point in time, and do not reflect the financial dynamics of that company. They do not, for example, take into account whether the company can actually finance its exploration programmes internally without dilution, or enhance its value by adroit portfolio management. The investor should also be interested in the likely future development of the company's asset value, and consequently the specific exposures and sensitivities that the composition of the assets (and management) provides. Figure 4.12 – showing the average consolidation of the asset breakdown for the basket of UK independent companies – illustrates the point that variations in the orientation of individual companies according to asset type imply different types of exposure to the investor. The figure shows that just over half of the current value of companies is contained in producing fields, and the majority of these will tend to be in decline with relatively high operating costs. However, these assets represent the engine room generating the cash flow to fund development and exploration work. However, the cash flow in most cases follows a high tax liability, which can be used to offset to varying degrees the exploration and development expenditure. A well-balanced and tax efficient portfolio will capture greater value from production, while a substantial imbalance may signal the possibility of enhancement through portfolio management or even a bid. The larger

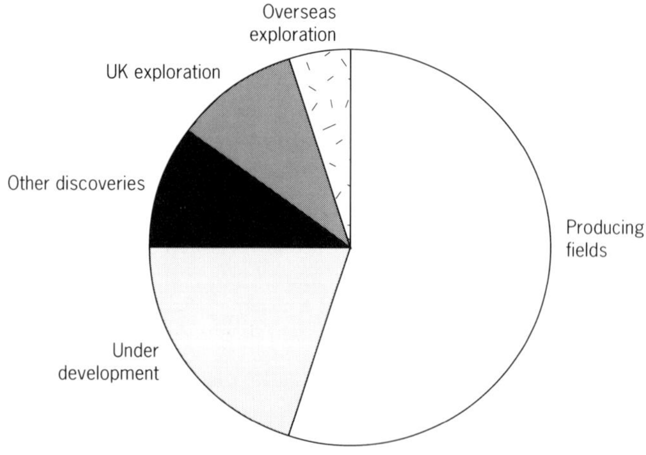

4.12 *Average consolidated value of E and P companies.*

the company, the harder it will be to maximise this tax efficiency. Nonetheless, this core of producing reserves is by definition in decline. It will offer better gearing to short term movements in hydrocarbon prices, but does not in itself represent gearing to longer term growth.

Fields under development represent the component of an asset value that underpins the growth potential of the entire group; they represent the established growth potential and enhance the value of the existing output. They offer exposure to the degree of risk in the development programme itself, reserves redeterminations upon gaining production experience, and longer term oil and gas prices.

Technical reserves represent the component of an asset where value appraisal is inadequate and commercially unproved. The value attributed to technical reserves is usually low, but can be inherently volatile. Although in some ways a lower quality asset which might involve expensive and ultimately abortive appraisal drilling, technical reserves can equally offer good exposure to advances in technology and a geared response to higher oil and gas prices if the minimum economic threshold is crossed. Until comparatively recently small non-satellite fields in the UKCS would have been perceived to have little or no value, yet advances in technology have made such fields economically attractive with reserves of only 8 mbbl.

The value of exploration acreage represents the final component of asset value, and is the most volatile. Some investors have difficulty in recognising the value attributed to this category despite the fact that it offers the potential to add value through exploration success. For the smaller company in particular, the acreage offers the possibility of bringing about a change in its fortunes, justifying its rationale. For the investor, a good acreage position and active drilling programme represent the best hope of adding value.

EVALUATING ASSET VALUE GROWTH

Core values and total values

The asset value of any company will not be accurately known until it has produced and sold its last drop of hydrocarbons from the last discovery in its portfolio, or unless it is the target of a successful bid. Until that time there will permanently be uncertainty surrounding each component of an asset value. The most uncertain value is the exploration acreage while the most certain are the producing assets. In order to make asset values more transparent it is necessary to look at the value of a company in terms of a 'core value' which is made up of the value of all its commercial and potentially commercial reserves and an 'upside value' which is made up of technical reserves and exploration acreage. Both of these values together make up the 'total value' of each company.

The core value comprises the commercial and potentially commercial reserves which drive the short term financial position of a company (cash-in from producing fields and cash-out for new developments). The upside value includes the technical reserves and exploration acreage which represent the future growth of the company, but which will also require a proportion of the cash flow in order to be realised organically. The relative proportion of core value to total asset values will vary from company to company depending on the quality of the acreage, as well as each firm's individual tax position.

Cash flow generation

So long as E and P companies are striving for real capital growth there is every incentive for them to maintain exploration expenditure, rather than return value in the form of dividends. This is certainly the strategy of most of the smaller E and P companies, and is reflected in the proportion of cash flow that is paid out in dividends. Obviously, as exploration companies mature, growth becomes harder to achieve through exploration success alone, and a greater proportion of cash flow should therefore be repaid to the shareholders. Furthermore, it could be argued that even small companies with relatively strong cash flows should return value to the shareholder if they cannot re-balance their portfolios with new acreage.

Cash flow is generated from the core value of a company and as such does not include any upside value that may arise from future exploration expenditure. The growth of the core value relies on reinvestment of this cash flow in exploration or acquisitions. Companies which invest most of their cash flow in exploration offer the greatest exposure to exploration success. However, if the exploration programmes are entirely unsuccessful it could result in erosion of the core asset value.

Exploration potential and finding costs

Gearing to exploration success is only one element to consider when looking at prospects for future growth. After all, there is little point in a company spending a large proportion of cash flow on exploration if the acreage portfolio is of limited prospectivity or contains prospects that may be uncommercial. Companies must strive to reduce finding costs in order to add significant value.

Finding costs can be defined as 'the ratio of capital expenditure on exploration and appraisal to the sum of the reserves discovered and extensions including reserves revisions'. The numbers are usually represented on a gross basis, but depending on the tax position of a particular company net exploration cost will probably be much lower, so leading to considerably lower finding costs. Exploration expenditure can to varying degrees be offset against tax or recovered from cost oil. These tax offsets obviously have an inherent value in an ongoing business, as long as it is realised that in order to obtain the relief, money has to be spent on exploration in the first place. Until April 1993 tax relief for exploration in the UKCS was as high as 83% of costs (a net cost of 17p in the pound), depending on the tax position of the company involved. However, with the abolition of tax relief against PRT in 1993 the net cost of exploration in the UK has quadrupled (to a net cost of 64p in the pound). Therefore, net finding costs are often much lower than gross finding costs, and this must be taken into account when evaluating the addition of value.

Calculated finding costs may be more pessimistic or more optimistic for different companies because of the way in which reserves are reported. For example, some companies report gross reserves, e.g. companies involved in production sharing contracts, whereas other companies report reserves net of royalty. In addition, since the average time between initial exploration, e.g. initial payments to host governments (signature bonuses), seismic surveys and the booking of reserves is between three and four years, companies which have been building up acreage positions may have unduly high finding costs in contrast to companies which have been actively testing prospects on an existing bank of prospective acreage. Companies may also hold back or carry forward discoveries from one year to the next, thereby distorting actual finding costs. Although there are clear limitations in the calculation of finding costs (without a consistent breakdown of annual reserves additions and expenditure from company to company) internal inconsistencies should average out over a period of time.

With prospective exploration budgets it is possible to calculate the reserves replacement ratios for an E and P company that will be achieved if the historical or average finding costs are maintained. Obviously those companies with the best historic finding costs are more than capable of replacing reserves if the record is sustained. In contrast, companies with below average finding costs may actually lose value.

It is highly unlikely that exceptionally good finding costs can be maintained for any length of time with a portfolio of acreage that is being actively tested. However, there is always the possibility of discovering the 'company maker', with the potential to transform a company overnight. In addition to finding costs it is also worth putting into perspective the actual volume of production that is required to be replaced. Companies with very high levels of production are unlikely to be able to replace reserves with one discovery alone, and must rely on a number of small incremental discoveries; in contrast companies with lower levels of production may more than replace production on the back of one discovery alone.

Core value growth

Although the application and projection of a reserves replacement number for an E and P company is useful, the value of new reserves to be discovered is of greater importance. After all, if oil is discovered at a cost of $5/bbl and the new reserves are only worth $3/bbl, the core asset value will be eroded, not increased! Value will obviously vary depending on the size and location of the new reserves.

Using the forecasts of the exploration budget it is possible to predict the increase (or decrease) in real terms of the core asset value against the market for various finding costs. This analysis – applied to any company – will show the tremendous gearing to exploration success if finding costs can be kept low. However, it will also show that companies can actually lose significant value if finding costs are too high. Therefore, if companies discover oil at industry average gross costs it follows that they will add value, and core asset value will grow.

Those E and P companies which plan to spend the greatest proportion of their cash flows on exploration offer the greatest potential to grow their asset value. However, they will only add value if costs are kept low. By projecting the historic and average finding costs into the future it is possible to calculate the potential core values of each of the companies in five years' time, and establish whether this growth in value is in excess of the return a shareholder may obtain from the market.

ACQUISITIONS

All of the methods discussed above outline the way in which an E and P company can grow itself organically. However, history has shown that the fastest way in which most of the E and P companies have grown is through acquisitions, made either to boost production or replenish exploration portfolios. Acquisitions by E and P companies have taken the form of either hostile bids for other companies, or the acquisition of packages of assets. There is nothing wrong with expanding a company through an acquisition; the real question that should be asked at the time is whether it has actually added value.

Companies will only add real value through an acquisition if they can acquire the assets at a price below their own net present value; this may be possible if there is a gap between the value of the assets to the purchaser and to the vendor. However, the gap in value is likely to be quite small on producing assets, depending on different oil price scenarios, tariff and tax positions. Real value is more likely to be added where interpretations of prospects on exploration acreage differ significantly. For example, one company may consider a block to be unprospective, whereas another company may regard the same block as being highly prospective. Obviously, if the purchaser acquires the acreage and subsequently makes a large discovery then significant value will have been added. With exploration acreage, the real value of an acquisition can only be realised through time, and may not actually be apparent at the time of the deal.

Historically, exploration companies have been very acquisitive when they have strong underlying cash flows and they have real confidence about the future oil price. This combination of circumstances has brought about inflated purchase prices, and it is all too often the investor who has paid the price by subscribing to overpriced equity issues. Companies also pay the price for acquiring assets at too high a price, because when oil prices fall or remain flat nominally they are left with a high depreciation charge which puts pressure on earnings – even though underlying cash flow might still be quite strong.

EXPLORATION COMPANY PERFORMANCE

Asset values, asset values' components and the potential growth of asset values are the main benchmarks that are used to value the E and P sector. However, asset values do not provide the investor with a guide to the appropriate premium or discount to asset value that he should pay for. This must come from a combination of performance indicators, financial strength and company strategy. Quantitative performance indicators include historical finding costs and reserves replacement ratios, and qualitative indicators include an assessment of management and company strategy.

Performance indicators

The main quantitative performance indicators are:

- Historical finding costs.
- Reserves replacement ratios.
- Asset value growth.
- Financial strength.

Each of these indicators is based on historical performance, and may not actually represent the future performance of a company if it has undergone a major acquisition. However, investors not only use these indicators to assess the future performance of a stock, but can also relate them to the more qualitative indicators, such as:

- Management.
- Strategy.

If a company has had a poor track record of exploration, and its reserves replacement record is poor as a result, its stock will usually trade at a deeper discount to asset value than a company which adds value with low finding costs. The quantitative performance indicators are often reflected in investors' and analysts' views on the quality of management and company strategy which may in turn exacerbate the poor rating of a company.

Company rating

Although performance indicators enable investors to judge the relative performance of one company against another, they do not give a firm indication of the appropriate absolute premium or discount to asset value of any particular company. However, there are two natural benchmarks against which shares can be judged. The upper benchmark is the level where there is no premium or discount to asset value. If shares are trading at a premium to this benchmark then they can be considered to be overvalued, since they are priced above a level that industry would be prepared to pay. Clearly, if shares trade above this level, the market must have a higher expectation of performance based on the indicators detailed here. The lower benchmark is the level where the stock is trading at such a discount to asset value that it would be compelling for a potential 'predator', who would find it cheaper to replace reserves through acquisition rather than through exploration alone. In between these two levels, the appropriate premium of discount is really only a function of the internal rate of return that each company is likely to achieve on its retained cash flow.

Since the core value represents the net present value of a company's free cash flows, with no gain or loss attributed to any exploration expenditure, it follows that if the company were to liquidate, shareholders could receive the core value in the financially inefficient form of dividends (less unrecoverable advanced corporation tax). However, all E and P companies are ongoing businesses involved in active exploration, and the ability to add to their core value depends on future exploration success. Therefore, if companies maintain their good track record of exploration they will justify a premium rating. Companies which continue with a poor record of exploration which looks likely to continue with its portfolio of exploration acreage should trade at a discount to its core value. In fact, if investors fear a programme of exploration will be unsuccessful there is little point in them continuing to invest in the stock, and they should sell immediately (liquidating their own holding) to pre-empt further erosion of the core value. At any particular time, there will be anomalies of one form or another in the relative rating of particular stocks.

The internal rate of return that investors require can be obtained by calculating the discount rate required for the share price to be equal to the core asset value. This value represents the internal rate of return that investors expect companies to generate from the core value based on the current share price. Figure 4.13 (and other similar charts) shows that the rate of return expected from companies with poor track records for exploration is often much higher than the rate of return expected from companies which have proven good track records. Such a figure may be more useful in judging the correct rating of a stock than a table of discounts to asset value as the internal rate of return can be viewed directly against the cost of its own equity. Stocks which appear to offer a rate of return lower than their cost of equity can be considered to be overpriced in contrast to stocks which offer a premium to their cost of equity.

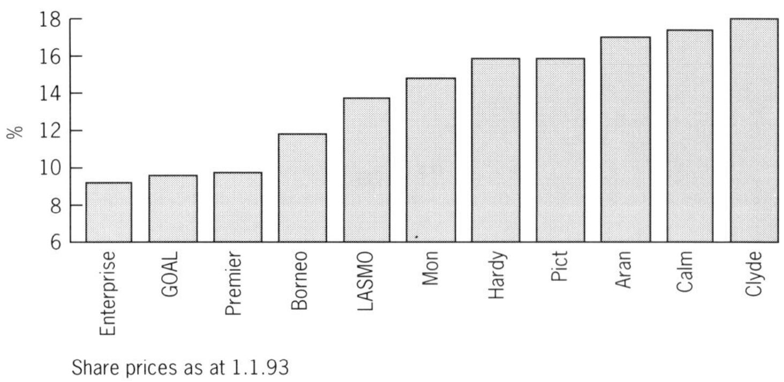

4.13 *Internal rate offered by commercial reserves.*

Conclusion

Ultimately, each exploration and production company is unique. A company might be very small, yet be aiming for growth through an aggressive exploration programme. Another may have steady cash flow with little upside potential from new development; it may also have reached a size where growth from exploration success will not have a material impact on the company (in which case other financial considerations such as dividend yield will begin to play a more important role). The character of a company must therefore be taken into account when investing in this sector, and any investment must of course coincide with the investor's own strategy and risk profile.

CHAPTER FIVE

Integrated oil companies

INTRODUCTION

Indirect ownership

Investors do not buy assets directly when they buy shares in a company. Rather, they buy an interest in the company itself, and through it they obtain an indirect interest in its assets. Unless the company is acquired, in which case they will receive a consideration for their shares, they will only receive from the company such of its cash flow that it pays out in dividend – as opposed to that retained as cash or reinvested in new capital expenditure.

Asset values of integrated companies

In the case of exploration and production companies, it is often a fairly easy matter to obtain adequate information to produce an asset value, while the lumpiness of annual earnings and cash flows usually make these largely useless as evaluative tools. The investment issue is then how to establish the appropriate premium or discount to asset value for any individual company.

The opposite tends to apply to integrated companies. These are groups of vertically integrated businesses, whose activities usually extend from the wellhead to the petrol pump, and often beyond the oil industry into petrochemical activities. In many cases, they also include non-petroliferous activities, and they are usually highly internationally diversified. Royal Dutch/Shell, Exxon and BP are among the world's largest – and most complicated – companies. To value them literally asset by asset would be virtually impossible – not just because of the scale of the task, but also because modelling the cash flows of an ethylene cracker (the plant which transforms petroleum product into the basic unit of plastics) is hugely more uncertain than modelling the cash flows of an oil field.

Fortunately, because the downstream (refining, marketing and chemicals) business activities of an oil company represent a mass of relatively small investment decisions, rather than a few enormous projects, the alternative methods of analysis of accounts and of calculation of historical cost returns on capital employed work better than for the upstream. In any case, such companies are too large – and often too politically important – for their acquisition to be acceptable to the authorities, even if it were practicable. Their asset value is never likely to be realised by a bid from another company, and the shareholder will almost certainly receive his return in the form of a stream of dividends. All the more reason, therefore, to value the shares not in relation to an asset value but in relation to the stream of earnings, cash flows and dividends which they will generate.

Investment theory

There is a body of investment theory on the evaluation of equities which essentially concludes that the two alternative approaches are to value the net free cash flows which a company will generate over time (prior to dividend payments and incremental capital expenditure), or to value the stream of dividend payments which an investor will receive from the company over time. In an efficient capital market with no fiscal distortions (highly restrictive conditions, it should be noted), the two approaches will come to the same thing, i.e. the investor will be indifferent to whether a company reinvests cash flows, or pays them out if its return on capital is the same as his discount rate.

There are a number of excellent textbooks which explain the economic theory surrounding the general question of the evaluation of shares, e.g. *Principles of corporate finance*, Brealey and Myers, McGraw-Hill). For the purposes of this book, which is specifically concerned with the application of these techniques to practical examples in the oil industry, we shall content ourselves with the conclusion that while the discounting of free cash flows was usually the most appropriate method for evaluation of exploration companies, it is the prospective stream of dividends which represents the most convenient measure of value for the equity of integrated oil companies. How this stream is discounted, however, is a function of the risks associated with it, so it is important to establish sensitivity analyses, as well as the base case.

Capital appreciation

A particularly important point which we must note here is that the use of the concept of the future stream of dividend as the basis of equity evaluation is not to ignore the element of capital appreciation in shareholder return. If the share price rises over a period it is because the present value of the stream of dividend which it offers is higher at the end of the period than it was at the beginning. Capital growth is implied by a rising dividend stream, but the present value of a share is simply the discounted value of the dividends which it offers. A share can thus be seen as a perpetual indexed annuity, about which the key question is the rate at which the income stream will grow, and the risks involved.

ACTIVITIES AND STRUCTURE

Introduction

Before moving on to methods of analysis used it would be useful to run through the activities and structure of an integrated company, since analysis of the prospects of a business does require that the analyst knows what it does, and is aware of the factors which can be expected to establish its profitability. Integrated oil companies are defined as those which have both upstream (exploration and production) and downstream (refining and marketing) businesses.

Integration

It is not, in practice, inevitable that a firm's upstream and downstream activities will be physically integrated in the sense that it puts its equity crude oil through its own refineries. Prior to its acquisition by LASMO in 1991, Ultramar was an integrated oil company, but its refineries were in Los Angeles and Quebec, and its production mainly consisted of natural gas in Indonesia (which was sold to a consortium of Japanese electricity utilities) and a portfolio of oil and gas assets in the UK North Sea (much of which had still to come into production at all).

In many cases, integrated oil companies are highly internationally diversified, though this is less true of those which were originally state refining companies, and which have only subsequently been privatised and diversified geographically. Of course, some of the state-owned production companies in OPEC and elsewhere have diversified downstream overseas, but here we are solely concerned with quoted companies.

State ownership

The issue of state ownership is still important for investment analysis of the quoted oil companies, however. As was pointed out in Chapter 1, the oil industry is characterised by the extent to which its assets, both upstream and downstream, are owned by governments, rather than by the private sector. An obvious consequence of this is the role of OPEC in influencing crude oil prices; a second consequence is the restriction on oil companies investing directly in upstream assets in the Middle East and Latin America. Thirdly, government control, even if not direct ownership, has frequently restricted downstream oil margins, as companies have been prevented from passing on oil price increases to their customers.

Diversification

In addition to vertical integration in pure oil activities, most integrated oil companies have diversified out of the oil industry to some extent. In most cases, this involves a further downstream extension into petrochemicals (which use oil products and natural gas as feed stock). There are many examples of oil companies diversifying from oil altogether. Many of these moves were made in the 1970s, under pressure from the then prevailing opinion that oil, as a non-renewable resource, was in danger of running out, which would obviously have meant that the companies would benefit from shifting their balance of activities into other industries. Given their phenomenal financial strength at that time, it is salutary to remember that these excursions into mining, nuclear engineering, coal, geothermal energy, power generation, speciality chemicals, fertiliser manufacture, forestry, food manufacturing, metals, retail, office equipment and, in one memorable instance, ownership of a circus, almost all proved disastrous investments for the oil companies concerned.

Many of the oil majors have largely or partly withdrawn from the activities into which they diversified in the 1970s, and while the residual involvements must of course be taken into account in analysing the individual companies, we shall concentrate on those activities which are typical to the group of companies, vertically integrated into one another.

EXPLORATION AND PRODUCTION

Introduction

The exploration and production operations of integrated companies are essentially identical with those of the pure upstream companies, except to the extent that they are usually much larger (this is not inevitable; Enterprise, LASMO and Oryx, for example, are all pure upstream companies but are considerably larger than the upstream businesses of small integrated companies, such as OMV, or many of the smaller US integrateds), so most of what has been covered in the previous section of this book will also apply to them.

Balance of portfolio

There are two scenarios in which the exploration and production operations of integrated companies differ from their pure counterparts upstream. The first is that an integrated company, even if it is not refining its own crude oil, will often have a target as to the balance between its upstream and its downstream operations, and it may be unwilling to permit the relationship between the volume of crude oil which it refines and the oil equivalent amount of hydrocarbons which it produces to move outside certain limits. This issue is contentious, since the balance which each company has attained is a legacy of its history, and if it is being managed to maximise its value it can be argued that no particular balance is necessary. The problem with this conclusion is that it ignores the risk associated with a different balance of activity regarding earnings volatility and sensitivity to price and margin changes.

A recent example of the dangers associated with excessive dependence on profits from a highly volatile market segment is the Belgian company Petrofina, whose earnings collapsed with the downturn in the petrochemicals market in 1992. If a company is to have a large exposure to such an industry, it helps to have a stable source of adequate cash flow from other activities.

A refiner, for instance, which produces significantly less crude oil per day than it refines, is exposed to the influence on refining margins of a surge in the crude oil price. Balanced operations will protect him against this risk, though with operations and accounts becoming more transparent the downturn in the refining contribution will still be apparent to investors. Moreover, with futures markets becoming more sophisticated, it is increasingly possible for companies to hedge elements of the risk associated with a particular operating imbalance.

Such strategic considerations apply more to integrated companies than to exploration companies, which can be more opportunist about their choice of capital investments (though even they have to worry about the timing of cash flows and how to finance themselves).

Natural gas

The second main difference between integrated and pure upstream companies is with regard to natural gas. Exploration companies tend to aim to sell gas at the wellhead. The oil majors are prepared to undertake the large capital expenditures required to build gas liquefaction facilities required for the export of liquefied natural gas (LNG) and its transportation. They are also increasingly entering into the transmission, distribution and marketing of natural gas into consumer markets where deregulation of the gas industry permits this.

Gas was, in its early years, regarded as a useless by-product of oil, and was flared off. When gas markets did develop, it was separate gas companies which built the pipelines through which it is distributed. The introduction of LNG created new investment opportunities for the oil majors, of which the most committed to this new market is probably Shell, followed by Mobil and Total.

Deregulation is extending the process into the pipeline industry. While these are properly speaking downstream integrations (for gas rather than oil), they remain closely associated with the exploration business of many companies. This highlights an essential difference between crude oil and gas. Oil, once produced, is cheap to transport, and once in a tanker is saleable anywhere. Consequently, the crude oil market is an international one, and oil developments are screened with respect to a world price.

Gas is the opposite. Its development cannot be planned without consideration of its transport to consumers, whether by pipeline or liquefied. The capital cost of both options is high, whether it is paid for by the producer or the buyer. It follows that gas is generally developed by large companies under long term contracts. Deregulation of the US gas industry has only been possible because of the extensive US pipeline system, and it is instructive that deregulation of the UK gas market is being forced by legislation, rather than evolving naturally. There was little pressure from the oil companies to encourage it to occur naturally.

Size offers stability

Other than for the two factors which differentiate the two, however, the exploration and production businesses of the integrated companies are similar to those of the pure exploration companies, though generally larger and more mature. This permits that the analysis of them can concentrate more on performance over time and profit potential, and less on asset evaluation by project, since the earnings and cash flows are smoother, and the finding costs and reserve replacement performance is over a large range of wells and projects, rather than reflecting a statistically unpredictable small sample of interests.

DOWNSTREAM OIL

Introduction

Although oil companies usually have a range of business streams occupied with downstream oil functions (shipping, supply and trading, refining and marketing, etc) external analysis tends to bundle these activities together. The main reasons for this are that they are usually accounted for under one heading in consolidated reports and accounts. The analyst may be aware of the different activities behind the quoted figures, but it is extremely difficult to obtain accurate data about them.

Downstream oil activity is competitive in a way that upstream oil is not. The companies do not usually operate in partnership (though Chevron and Texaco's long-standing Caltex joint venture in the Far East is an exception), at least in their marketing activities. They compete for market share; commercial information about sales volumes, cost structures and even contract prices are jealously guarded secrets. Moreover, it is the downstream end of an oil company's activities which attracts the political opprobrium, a classic example of shooting the messenger, since examination of oil company accounts makes it clear that it is generally the upstream business which earns good returns on capital (and upstream prices are outside the control of the companies).

Where joint ventures downstream are more common is in the ownership of refineries, and more recently in the joint ventures which have been established between OPEC states – notably Venezuela and Saudi Arabia – and Western oil companies. The latter is in effect a move back towards the vertical integration which was eliminated at the time when the states concerned nationalised their crude oil reserves, leaving the companies with a preponderance of downstream assets.

Shipping

Since the 1970s, companies have struggled to generate a respectable return on capital from their refining and marketing operations (in fact, it has been refining which has caused the trouble). Of the downstream businesses mentioned above, the oil companies have substantially reduced their exposure to one – namely shipping – after their heavy losses in this business during the 1970s. Although the ageing of the tanker fleet represents a risk to the oil companies (and some are now again expanding their tanker fleets) they now for the most part contract out this function to third party shipping companies.

Supply and trading

The second downstream business, supply and trading, represents the commercial function of deciding how to manage the sale of a company's upstream production, how to source the supply of crude oil and feed stock to its refineries, and how to manage the transfer pricing between the refineries and the marketing operations.

Over the past twenty years or so, the replacement of fixed price long term contracts with spot prices and the use of futures contracts to hedge price risk has transformed this function in oil companies. Increasingly, companies have made internal transactions more transparent and 'arm's length', even where they have not been required to do so by fiscal authorities. Removal of long term supply arrangements has resulted in refiners having an incentive to widen as far as possible the flexibility of their plant in terms of their ability to use alternative feed stocks, although this always carries the danger that price differentials will not in fact justify some of the investment involved.

For the supply function technical logistics have been largely superseded by trading considerations. Supply departments employ few fixed assets. Their activities are indistinguishable, to the outside observer, from those of the refiners and marketers. Crucial though their activities are to each company as a whole, they are an unquantifiable asset to the shareholder. As with other managerial functions, their existence is of no significance whatsoever to the break-up valuation of a company. Their contribution to a company's performance over time, however, can be considerable, and will merely represent itself as one of the key factors causing it to generate a relatively higher or lower return on its capital employed than its competitors.

REFINING

The industry

Introduction

If supply is a largely managerial function, then the opposite can be said for the refining business. This is where the heavy downstream capital expenditure is allocated, and the versatility, profitability and environmental friendliness of a plant is a direct function of its sophistication and its age. In essence, a refinery distils crude oil into a range of petroleum products, which are then marketed to end users.

An excellent description of the technicalities of refining is available in the relevant chapters of BP's standard textbook on the oil industry, *Our industry petroleum.* Since some knowledge of what happens in a refinery is necessary if sense is to be made of the analytical methods to which we shall turn later in this chapter, however, we provide an outline explanation here.

Refining process

Crude oil is a mix of hydrocarbon compounds of different molecular weights. Hydrocarbons begin with methane, natural gas, and then extend through ethane, propane and butane (liquefied petroleum gases), followed by a vast range of compounds which range from very light, volatile fractions to, at the heavy end, bitumen.

The two most important characteristics of a crude oil to a refiner are how light or heavy it is (that is, whether it contains greater proportions of the lighter or heavier compounds) and how much sulphur it contains. Sulphur is a nuisance as it reacts to create corrosive compounds and is therefore removed in the refinery. Environmental legislation is progressively reducing the permissible amount of sulphur in oil products.

The crude oil distillation unit is the heart of any refinery (in the early days it was the refinery), and it distils the crude and separates it into light distillates, middle distillates and residual fuel (that which doesn't boil in the primary distillation unit at atmospheric pressure). End user demand for petroleum products broadly follows this characterisation by relative density.

Refined products

Gasoline and naphtha (petrochemical feed stock) are both light distillates. Aviation kerosene, diesel engine oil and domestic heating oil are all middle distillates. Residual fuel has a variety of applications, but it can be burned directly in ships' boilers, in electricity generation plant or used as heavy fuel oil by industry. Thus, apart from the speciality products (lubricants, wax, bitumen, etc), a simple distillation unit produces all of the main categories of oil product.

Upgrading

The other units in a refinery are employed, for the most part, to do one or more of three things:

- Convert the heavy end of the product barrel, residual fuel, into additional quantities of distillate.
- Reduce the sulphur content of the oil products produced.
- Improve the combustion characteristics (the octane rating) of the gasoline which the refinery produces.

Demand trends

As we have discussed in the section on oil pricing, there has been a tendency for demand for oil to shift towards a lighter and lighter product slate, as demand for transport fuels grows faster than that for heavy fuel oil. This alone would require refiners to add conversion plant. At the same time, however, refiners have become less willing to commit themselves to the use of particular crude oils, as this exposes them to price risk in volatile crude oil markets. In addition, they are faced with the prospect that a greater proportion of their supply will probably come from the Middle East, whose crude oils are in general heavier than those from the mature North Sea, for example.

Conversion plant

Conversion plant is amongst the most expensive, and the most important in a modern refinery. The methods used by different refining processes used in conversion vary considerably. The oldest and simplest involve simple thermal cracking. Probably the most frequently used are catalytic crackers. And there is now an increasing interest in 'deeper' conversion units, such as delayed cokers and hydrogen conversion units. These processes are more radical than the earlier, simpler upgrading processes, and can often use any quality of residual fuel as feed stock. This contrasts with the older plant, for which the conventional configuration is for residue from primary distillation to be vacuum distilled, and for the cracker to use as feed stock the vacuum distillate from this second distillation process.

Desulphurisation

The term desulphurisation is fairly self-explanatory, and the process is environmentally driven. Again, the plant is expensive, and does not add directly to the market value of the product produced, except to the extent that there is likely to be less capacity to supply the higher quality product. This is increasingly a dilemma for refiners, that they are being required to add expensive capacity for environmental reasons, with the only prospect of recovering their costs being the potentially tighter supply capacity in the market for the higher quality product.

Octane

Finally, there is the issue of the octane pool. The light distillates which are produced from primary distillation often have fairly poor combustion characteristics. As internal combustion engines have become more sophisticated, this has become more of a problem. The cheapest solution for refiners was to add lead to the fuel, and this was the preferred solution until environmental pressures resulted in its phasing out. An alternative process, the reformation of light distillates to produce higher octane product is now the most popular process, but it too has run foul of environmental legislation. In the US, where the Clean Air Act has effectively imposed a chemical composition for gasoline on refiners (so-called reformulated gasoline), the amount of benzene in gasoline is limited to well below that which would typically be contained if reformation were the main process used; refiners are required to use increasing quantities of oxygenate additives (notably MTBE) as yet another way to raise the octane rating of gasoline (other more expensive alternatives to reformation are isomerisation and alkylation).

As with sulphur extraction, the key point for the industry is that while it is suffering from an increase in its costs, its products are no more effective from the point of view of the consumer. Cost recovery will depend on higher capacity utilisation, and the extent to which the industry can raise prices. Clearly, when evaluating individual companies, the complexity and state of their refining plant is the key to valuation of this business.

Blending

The description of the refining process given here probably conveys little of the complexity of the blending and unit optimisation involved. Refiners vary the crudes which they process to adjust for cost and for their product requirements, and they often have choices as to which combination of crudes and feed stocks to run to minimise the cost of producing a particular product slate. Linear programming models are used to permit this optimisation, and the refining configuration generally permits considerable blending of products throughout the refining process.

Figure 5.1 illustrates this point, showing the configuration of the OMV refinery at Vienna (the only refinery in Austria, producing some three quarters of the country's needs, and therefore inevitably required to be fairly sophisticated in its ability to match changing seasonal demand patterns).

Investment implications

While it is unusual, in practice, for an external analyst to be able to obtain adequate information to be in a position to value a refinery by detailed modelling of its individual components, it is necessary to be able to assess the overall complexity of the plant, and to be aware of the benefits and problems associated with a particular refinery configuration.

For example, if a refiner is entirely dependent on reformation plant to produce high octane unleaded gasoline, it becomes vulnerable to restrictions on benzene content, of the kind which has been introduced in the US. The larger units in a typical refinery and their functions are briefly described in the following section. As we have seen, how they are configured and which of a range of alternatives is adopted will be dictated by economic forces, by environmental legislation and not least by what plant the refiner already has available to him. There are few grass root refining projects outside the Middle and Far East. Western refiners usually have to bolt new plant on to existing units.

Refining plant

The main items of plant are as follows:

- *Crude distillation unit.* This is the basic unit which distils crude oil and separates out the main fractions of distillate from residual fuel. When the capacity of a refinery is referred to it is almost invariably the scale of this unit which is implied. Very simple refineries contain little more than this unit and some desulphurisation facilities.
- *Vacuum distillation unit.* This unit distils the residue which did not boil at atmospheric pressure in the crude distillation unit. It produces a heavy distillate, vacuum gas oil, and a bitumen residue, which after treating can be sold as bitumen. There are two main uses for vacuum gas oil. It is the standard feed stock for conversion plant to upgrade into distillates (most conversion units cannot be fed with untreated residual fuel, especially if the original crude oil was not light and low in sulphur), and, once treated, it is used as base oil which is blended with additives to produce lubricating oils.
- *Hydrofiner.* Sulphur is a recurrent theme in any discussion of refining plant, and the permissible level of sulphur in oil products is steadily being reduced, as environmental tolerance of such problems as acid rain decreases. Hydrogen treatment of products at various points in the refining process, prior to conversion or to final blending, is thus increasingly a necessity.
- *Visbreaker.* The simplest conversion processes are those which rely on heat to crack the hydrocarbon molecules. It results in a lighter product barrel than could be achieved by simple distillation. The light products which result from simple thermal cracking are of inadequate quality for modern requirements, and all require hydrogenation. One thermal process still in general use is visbreaking, which reduces the viscosity of residual fuel, creating a saleable heavy fuel oil without requiring the residue to be blended with gasoil or kerosenes to reduce viscosity.
- *Coker.* Also a thermal process, coking is aimed at eliminating the heavy end of the barrel. It upgrades residual fuel to a range of distillates, leaving a residue of saleable petroleum coke. A coking refinery thus eliminates any production of the low value heavy fuel oil product, though it is relatively expensive to operate. As with other thermal cracking processes, the gas oil, or naphtha, produced requires hydrogenation.

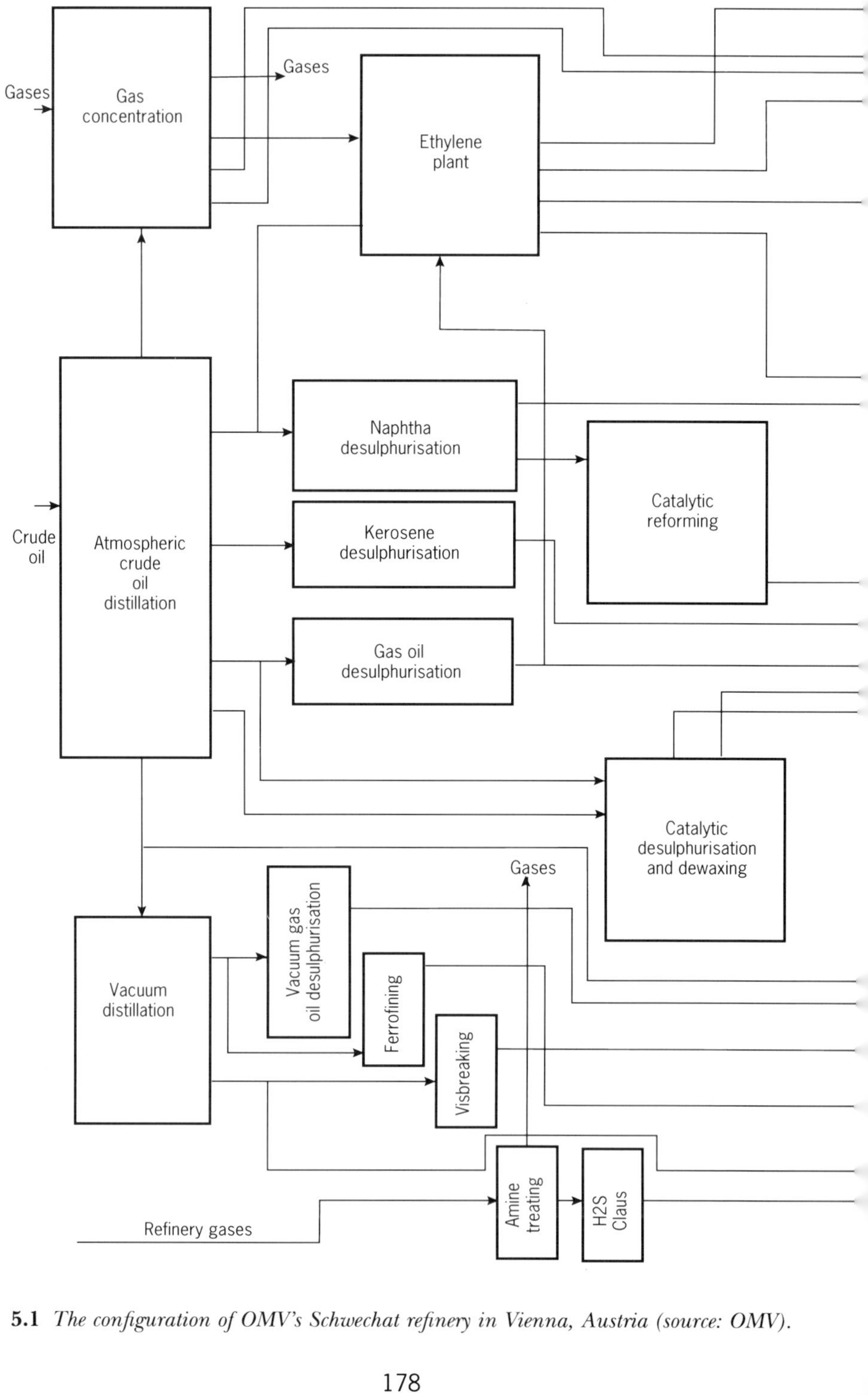

5.1 *The configuration of OMV's Schwechat refinery in Vienna, Austria (source: OMV).*

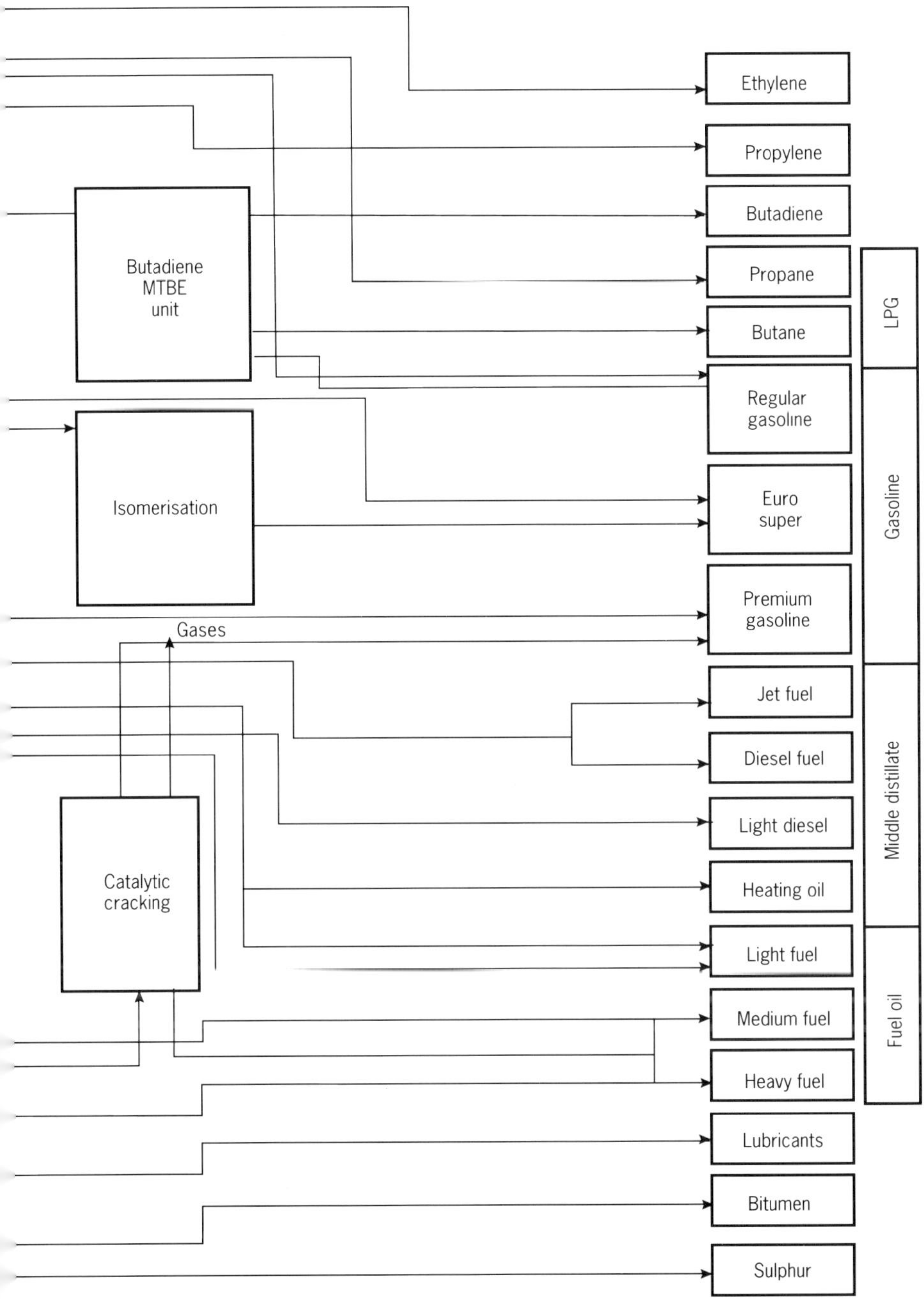
Ethylene
Propylene
Butadiene
Butadiene MTBE unit
Propane
Butane
LPG
Regular gasoline
Isomerisation
Euro super
Gasoline
Premium gasoline
Gases
Jet fuel
Diesel fuel
Light diesel
Middle distillate
Heating oil
Catalytic cracking
Light fuel
Medium fuel
Fuel oil
Heavy fuel
Lubricants
Bitumen
Sulphur

- *Catalytic cracker.* The use of catalysts can reduce the heat required, and therefore the operating cost, in the cracking process. It can also produce higher value product. Cracking of vacuum gas oil into distillates, for example, produces a relatively high octane yield of gasoline by comparison with simple distillation.

- *Hydrogen conversion unit.* At the cutting edge of refining technology these very expensive processes can run almost any feed stock, however heavy, bituminous and high in sulphur, and crack it into a relatively light slate of distillates. These units are exceptionally expensive to construct.

- *Hydrocracker.* An alternative approach to thermal or catalytic cracking, hydrocracking is an extremely flexible method for either upgrading vacuum gas oil into a slate of gasoline and middle distillates, or for simply upgrading gasoil to gasoline. The unit requires large amounts of hydrogen, however, and operates at high pressures and temperatures.

- *Reformation unit.* Gasoline is not merely defined by its molecular weight, but also by its combustion characteristics. In the absence of lead additive, the cheapest process designed to improve the combustion characteristics of gasoline is reformation, by which a catalytic process reforms paraffinic compounds into olefins, thereby enhancing the octane rating. Reformation uses naphtha as its feedstock, and produces high octane gasoline.

- *Isomerisation unit.* Reformation is an efficient method for producing high octane gasoline from naphtha, which has a wider range of molecular weights than gasoline; gasoline derived from distillation and thermal cracking is often poorer. Isomerisation, which alters straight chain paraffins into branched chain paraffins, offers an alternative approach to expanding the octane pool. It also avoids the problem of the high proportion of benzene in reformed gasoline.

- *Alkylation unit.* This process involves the polymerisation of propane and butane (LPGs) into iso-octane, which is the standard used for octane number. It offers an alternative use for LPGs, other than for fuel and for petrochemical feed stock, and it again avoids the problem of excessive reliance on reformation.

MARKETING

Introduction

The marketing operations of an integrated oil company are most widely recognisable in the shape of chains of service stations, but involve all of the products which are supplied to end users. There are wholesale oil product markets (the so called Rotterdam spot market, for example) in which it is possible to buy barge cargoes of oil products from refiners, and sell them on to end users or independent retailers of gasoline. Thus, there is in commercial terms a clear break between the refiner, who purchases crude oil and sells products, and the marketer, who purchases products at a wholesale price and markets them in inland markets at a retail price.

This distinction applies as much in the integrated companies as for the independents, as the majors use spot wholesale prices as the basis for their internal transfers; their marketing operations are required to generate a margin over the transfer price. Most petroleum product marketing involves selling to commercial buyers, for whom the determining factors are cost and whether or not the product meets required specifications.

Customers

The important commercial customer groups for main products are shippers for bunker fuel, airlines for aviation kerosene, electricity generators for fuel oil and industrial buyers of both fuel oil and diesel fuel. Commercial customers do also buy speciality oil products, such as lubricating oils, waxes and bitumen, for which product quality and (often more important) technical support count for more, rather than being entirely price driven.

Yet the customer for whom price is not the sole determining factor, and to whom the oil majors devote the largest part of their marketing capital and effort, is the retail customer, and in particular his purchases of gasoline, diesel fuel and automotive lubricant for his car.

There is a retail demand for heating fuel and in some countries, such as Germany, this is a large part of the demand barrel, though it is again largely driven by price. The attraction of the retail gasoline market to the marketer is that it offers the opportunity to market on the basis not only of price but also of service, and to a customer base which is susceptible to advertising and promotional schemes.

Gasoline retailing

Gasoline retailing can be and often is carried out by independent chains. Indeed, of the branded sites which comprise each major's own chain, some may not even be owned by or operated by the given companies. The trend is not to company ownership but to dealer operation under franchise arrangement, although this is not universal.

In addition to the supply of gasolines and diesel fuel at these retail outlets, the oil companies make a significant and increasing proportion of their profits from retailing other services at convenience stores, from catering (especially at motorway service stations), and from other services, such as car washes. A not insignificant factor is the capital gain that they can hope to make on their property.

The driving factor behind site throughput volume is location, and this is the key to the successful creation of a gasoline retail chain. Private individuals will tolerate a small price premium for the convenience of not having to make a detour to fill up their cars. In a low margin business, such as petrol retailing, this premium is the key to relatively high profitability. Moreover, prime locations offer higher throughputs, and sufficiently high volumes can offer good returns on capital, even given low prices. This has been illustrated by, for example, Atlantic Richfield's marketing operations on the US west coast, an isolated market in which it has a dominant market share.

DOWNSTREAM PROFITABILITY

Introduction

Many of the caveats which apply to the use of historical cost accounts to assess the performance and value of upstream businesses do not apply to the downstream. These operations do not generate their revenue by depleting their fixed assets. Refining is a manufacturing business to the extent that it requires the application of fixed assets to transform raw material into finished products. Marketing is a conventional retailing operation in that the margin between the wholesale price and the retail price arises because of the need for a distribution system.

Inflation means that the surplus as derived in accounts arguably does not fully provide for maintenance of the assets, but this applies equally to all other businesses, and does not invalidate historical cost accounts as a guide to profit and value.

It remains the case that industry acquirers of individual downstream assets, or even packets of assets, will attempt to value them in the same way that we have seen applied to upstream assets (by modelling the capital expenditures which will be required to be spent on the assets and the revenues which they can be expected to produce), but this is much more uncertain in businesses with narrow and fluctuating margins than it is upstream.

The factors which drive profitability in the two downstream businesses of an integrated oil company are quite independent from one another, however, and we treat the two operations separately.

Refining profits

Introduction

A refinery generates its profit from the difference between the cost of a barrel of crude oil feed stock (or other feed stock if it is, for example, buying in residual fuel to upgrade) and the value of the slate of products which it produces from the barrel of oil. To be netted off against this gross margin are the cost of operating the refinery (mainly fuel and some staff cost) and the cost of shipping the crude oil from the point of purchase to the refinery.

There has been a historical tendency for refineries to be built close to consumer markets because it is cheaper to transport crude oil than finished products, though OPEC expansion plans and environmental legislation are beginning to reverse this.

Netback calculations

Table 5.1 illustrates a typical calculation made for a notional north-west European refinery running Arab Light crude oil to produce a particular product mix. In reality, most refineries are flexible in the crude oil slate that they run. They constantly change the product slate produced. They do not only run crude oil as a feed stock and sell only

Table 5.1 *Netback calculation for Arabian Light crude at Rotterdam*

Product		*Spot price, $/bbl*	*Yield, %*	*Value, $/bbl*
Naphtha		21.71	5.4	1.17
Premium gasoline		25.63	11.9	3.05
Regular gasoline		24.00	4.3	1.03
Gas oil		24.29	40.3	9.79
Fuel oil	1% sulphur	16.54		
	3.5% sulphur	11.70		
	Adjusted price	11.74	32.8	3.85
Gross product worth				18.89
Incremental refining cost				−0.30
Spot freight cost				−1.27
Implied FOB value of Arabian Light crude				17.32
Spot price of Arabian Light crude				−15.70
Refiners' netback margin for Arabian Light crude				1.62

Source: Petroleum Intelligence Weekly, *August 17 1992.*

finished products to customers or other oil companies. They do not only produce main products. At times of low margins in particular they may make a significant proportion of their profits from special products, e.g. lubricants and bitumen. 1992 was not a good year for the refining and marketing businesses of the international oil majors, but it is still impressive that Exxon, in its report and accounts, noted that 'lubricants and related specialities contributed almost 20% of downstream earnings in 1992'.

In addition, in a low margin business like refining small differences in operating costs or efficiencies in terms of down time for maintenance or unscheduled closure, will make a decisive difference between the profitability of different units.

Applications

So, is analysis of standard margins of the kind illustrated of any use whatsoever in evaluating companies? It is useful, though it clearly does not permit the evaluation of individual plant in the way that we can model the upstream. (One could, of course, model individual refineries on the basis of certain feed stock and product prices, but only if one had enough data regarding their individual cost structures.) What the standard margin approach does do is to help to relate historical and projected trends in prices to trends in profitability. And two points rapidly emerge:

1 The geographical differentiation of product markets has direct implications on the value of location to refining plant.

2 The value of upgrading plant is considerably affected by the extent of the spread between the prices of fuel oil and light distillates. In other words, the advantage enjoyed by the owner of a complex refinery is not even over time.

Margin histories

Geographical differentiation and returns to upgrading, as opposed to distillation, are illustrated by the refining margin histories shown for the Far East and the US Gulf (see Fig. 5.2). The main feature of the first is the effect of the destruction of the Kuwaiti export refineries by Iraq on Far Eastern refining margins in late 1990 and early 1991, resulting in uncertainty over future supplies. In the US, the spring rise in the gasoline price in spring 1989 and 1990 improved margins to a complex refiner, but the onset of recession and Presidential exhortation not to 'profiteer' from the Gulf crisis caused the collapse of summer 1990.

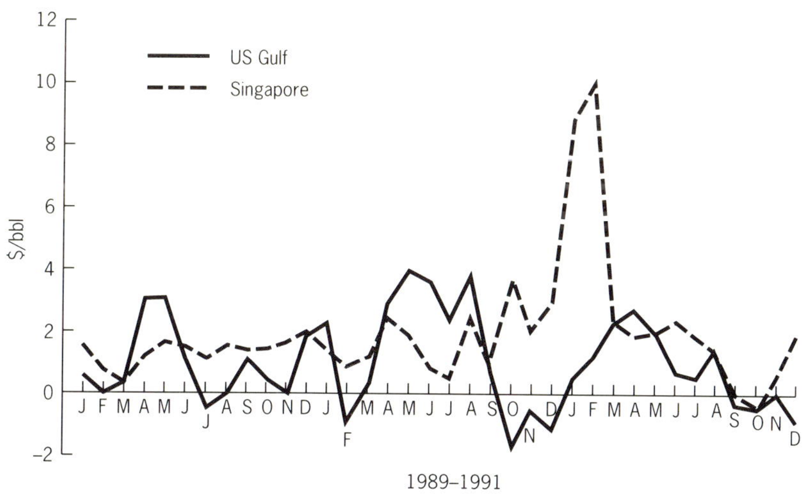

5.2 *Refining margin histories for the US Gulf and Singapore, 1989–91 (source:* Petroleum Market Intelligence*).*

Table 5.3 *Elf Aquitaine's French retail network*

	Number of gas stations	*Average monthly output per station, m³*
1990	4218	84
1991	2826	120
1992	1823	194

Source: Elf Aquitaine.

Table 5.4 *European hypermarket/supermarket share of gasoline market, 1990*

	1980	*1985*	*1990*	*1995*	*2000*
France	12.0	22.0	39.0	35.0	25.0
Germany	8.7	8.9	9.0	?	?
UK	1.1	3.0	8.0	16.0	25.0
Austria	–	2.6	2.2	2.2	2.2
Belgium	5.0	5.0	5.0	5.0	5.0
Finland	–	2.0	2.7	3.0	3.5
Greece	–	–	–	–	–
Ireland	–	–	–	–	–
Italy	–	–	0.3	0.6	1.5
Portugal	–	–	0.5	3.0	6.0
Spain	–	–	–	–	–
Switzerland	12.6	12.1	11.2	15.0	15.0

Source: The Institute of Petroleum, Petroleum Review, *June 1992.*

Site valuation

An industry buyer of an individual service station (there is an active market between oil companies in stations) will value it in terms of the cash flow which it can be projected to generate, as with any other asset. The overall portfolio will be assessed in terms of the commercial logic of its links with the company's distribution system, and whether the company has as a strategic aim to be a surplus or a deficit refiner. For example, in the early 1980s BP took the view (subsequently proved correct) that the European refining industry would remain unprofitable for some years, and that it would be considerably longer before it returned a commercial return on capital. However, BP also predicted that it would remain profitable enough to retail in most markets, and that the surplus of refining capacity would ensure that the deficit refiner suffered no serious disadvantage from having to buy in a proportion of its requirements. BP correspondingly closed down a large proportion of its European refining capacity.

The value of a service station is underpinned by its alternative value as a piece of real estate. This implies that companies will only build grass-roots sites when they can project that the profits generated from them will cover the cost of land and construction, and correspondingly that they will sell sites for non-oil use if the land value exceeds the value to them. (There is a cost to the latter course in that environmental remediation can be expensive, although not prohibitive.)

In the absence of precise accounts for the marketing operations alone, rather than the usual consolidated downstream figures, the two rules of thumb used in retailing evaluation are to take a reasonable average for the market in cents per litre of products sold (this will of course vary from market to market), or to use an average land value per site, which should understate the value of the business as a going concern.

Investment conclusions

For traders in the shares of integrated oil companies, rather than of the independent marketers, the profit contribution from marketing is the smaller and less volatile component of the earnings generated from the refining and marketing business, and of course the shares offer exposure to the combined entity. It is important to understand the marketing operations, therefore, not so much as to evaluate them individually, but to be able to see how they fit with a group's other operations, and how it compares with its competitors.

Having now described briefly the downstream oil operations of integrated companies and the factors which drive their profitability and value, we can return to the issue raised at the beginning of this section, the evaluation of integrated oil companies.

COMPARATIVE ANALYSIS

Introduction

Competitor analysis

There is a variety of reasons for analysing oil companies, and a variety of appropriate methods to use, depending on the requirement of the analyst. Oil companies increasingly analyse their own performance, as against that of their competitors, to establish their areas of comparative advantage and disadvantage, with the intention of trying to find ways of eliminating the latter. From time to time, they might also analyse other companies, or bundles of assets, from the point of view of a potential acquirer. This implies putting an absolute value on the company or assets concerned. Even in this instance, however, the value to the company of a collection of assets will probably be different to that of a competitor, and not only because of different macroeconomic assumptions. There would probably be tax synergies, or even the removal of exposure to a particular type of risk, such as feed stock cost, perhaps, for which the company concerned would be willing to pay, but which would be of no value to another bidder.

Investor appraisal

Investors also vary in their approach to valuation. There are in general two ways in which investors manage portfolios. The first is to screen equities for value and to buy shares which look underpriced, irrespective of the exposure which they offer. The second is to form a strategy based on opinions as to how the economic world is going to change (views on exchange rates, interest rates, economic growth, commodity prices, etc), and then to construct a share portfolio which is tilted appropriately. The two approaches both have implications for the appropriate methodology which should be used by the analyst. We shall start with the comparative.

The issues

Performance and balance

If it were the case that all oil companies had exactly the same balance between upstream and downstream activities (not to mention non-oil activities), and that they all had identical geographical exposure, then comparison would merely involve straightforward ranking by profitability.

In reality, of course, this is not the case. The companies offer differences of exposure, and at particular times will benefit from or suffer from the shape of their portfolio relative to that of other companies in the same industry. In other words, a comparative analysis has to address two questions:

1 How does the company's balances of operations compare with those of other integrated oil companies?

2 And how well does it perform in each of those areas of operation by comparison with other oil companies?

Financing
In addition to these questions about performance and ability can be added the question of how a company is financed (the proportion of capital employed represented by debt and by equity), and how its shares are rated in the market. The last point is not unambiguous. Ratings may vary depending on whether the comparison is based on earnings, dividends or cash flow.

Profitability
Oil companies themselves devote considerable effort to comparing their own performance with that of their competitors. Each, of course, will attempt to return the cost of capital or more, but from one year to the next its ability to do this will be affected by factors outside its control. Profitability targets are only meaningful in the long term. Comparison with other companies, however, is always applicable since a company's competitors will presumably have been operating in the same environment, and will have been subject to the same fluctuations in prices and demand (subject to the differences which are attributable to their differing portfolios of assets).

Profitability comparison

Return on capital
The simplest and most obvious form of comparative analysis is to compare the returns on capital achieved by the cohort of companies in the sector. Return on capital prior to financial charges and including debt and equity in the denominator is the best measure of operating performance, as it excludes the effect of balance sheet gearing on profitability.

Debt adjustment
In a good year, the return on equity of a highly indebted firm will tend to be higher than that of a similar, but unleveraged firm. In a bad year these roles reverse. Return on equity (earnings divided by shareholders' funds) therefore measures two things rather than one, though it is

return on equity over time which determines the return to the shareholder. (Over the short term the link between return on shareholder's funds and returns to the shareholder may diverge, owing to dividend policy or market movements, but over the long term they must converge.)

As most of the tax charge to an oil company is in the form of upstream taxes, not corporation tax, the usual method for calculating return on capital is to use after tax profit plus charges for net interest and minorities as the numerator, and debt plus equity (including minority interests) as the denominator, thus deriving a net of tax return on capital employed which is adjusted for the effect of balance sheet gearing. (A second reason for using net of tax returns is that many countries do not have a system of imputational tax credits, as in the UK, so the corporate tax charge is a deduction prior to gross returns to the equity investor.)

Balance sheet adjustment

After debt adjustment, there are still problems with the numbers as derived. UK oil companies state their earnings FIFO, whereas most US and continental European oil companies state them LIFO. It is therefore necessary to use LIFO figures for the earnings of the UK majors. These will still leave a small incompatibility in the figures for capital employed, since the working capital figure in the balance sheet will also be different for FIFO and LIFO accounters. One way to adjust for this is to adjust the US company balance sheets to reflect historical cost valuations. The resulting figures are inconsistent to the extent that they reflect LIFO income and historical cost capital employed, but at least they are fairly consistent between companies. Table 5.5 shows a series of returns on capital for a number of large oil groups over a five year period.

Table 5.5 *Oil majors' returns on capital, %*

	1987	*1988*	*1989*	*1990*	*1991*
BP*	12.1	10.2	13.2	12.7	5.9
Chevron	7.8	10.1	3.2	11.9	7.5
Exxon	11.9	14.0	9.7	12.8	12.8
Mobil	8.9	10.6	10.0	10.1	9.6
Shell*	11.0	11.5	13.7	12.2	8.4
Texaco	–	11.8	18.5	12.1	10.4

**FIFO accounts (US companies LIFO).*

Limitations to analysis

While it could be argued that profitability is ultimately the only thing that really matters to a business, return on capital alone does not get us very far, for two reasons. Firstly, there are huge distortions introduced by accounting techniques. Differences of depreciation policy and treatment of exploration expenditure distort both earnings and balance sheets. Old refining plant, often almost completely depreciated, will tend to generate high returns on capital, and so on.

Furthermore, consolidated figures do not distinguish between companies which are doing relatively well because they are exposed to parts of the industry which are booming, and those which are exposed to the same activities as their competitors but simply do them better. Pursuit of these issues inevitably requires subjective assessments of which criteria to compare, but results in much more useful analysis, and is the approach most generally favoured in the investment community. The list of potentially useful comparisons is endless, but the most frequently applied are listed in the following section.

Other analysis criteria

Upstream, it is reserve replacement success and finding cost which are most often compared, as these should provide guidance as to the long term trend in future production volumes and costs per barrel. Oil companies can do nothing to influence the price at which they sell their upstream production, though they can negotiate contract terms and taxation terms to some extent. In certain instances, such as in Abu Dhabi or Nigeria, they produce under fixed margin arrangements, and price movements can be almost or completely irrelevant.

Downstream, the information provided in reports and accounts is generally much less detailed, and the capital employed figures in particular are misleading since much of the refining plant dates back to the 1970s, appearing in books at virtually nothing. Analysis tends, therefore, to concentrate on the operational exposure, geographically and by balance between refining and marketing volumes, and on profitability measured not by return on capital but by profit per barrel refined (or sold, but this involves complications with profits or losses made on trading activities). All of these figures can be misleading, and none of them singly provides a definite sense of value, but in combination they are indicative.

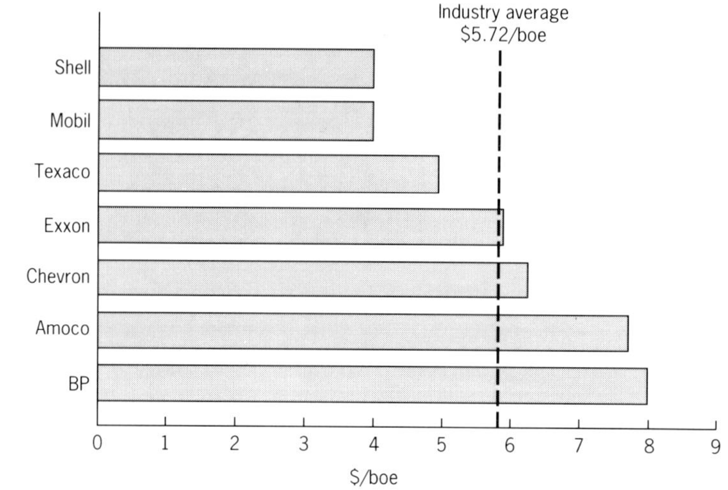

5.3 *Oil majors' oil and gas finding and development costs, 1985–89 (source: Shell).*

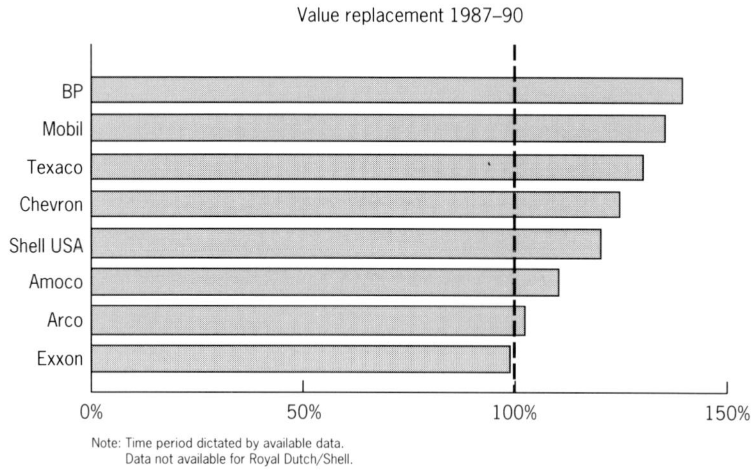

5.4 *Oil and gas value replacement, 1987–90 (source: BP).*

Reserve replacement

Proving anything

Figures 5.3 and 5.4 have been included to show how misleading comparisons can be; these figures were shown at presentations to investment analysts in London by the Royal Dutch/Shell Group and by BP within weeks of one another in the spring of 1991. Both are designed to illustrate the success of the groups as against their competitors in the area of reserve replacement. Each has managed to find a time period and a methodology which places itself at the top of the list, and it is also notable how little correlation there is between the ranking of the other companies listed. Neither methodology is entirely transparent, and we would recommend analysts to avoid both and adopt a simpler and more transparent alternative, even though the methodology behind both the charts shown can be justified.

Finding and development costs

Shell's approach recognises that the cost of bringing new oil and gas reserves on stream is not merely the cost of finding them, but also of bringing them into production. Figure 5.3 from the Shell presentation thus represents the result of calculations made by the US based consultant, J.S. Herold, as to the finding and development cost of new reserves to the group of companies concerned over the period 1985–1989.

Unfortunately, the reserve data presented by oil companies does not normally permit this data to be derived directly. It shows the movement in proved reserves attributable to discoveries, revisions, acquisitions and production, and the split of these between those which are developed and those which are undeveloped. However, in the event that the companies have bought or sold reserves, some of which were developed and some of which were undeveloped, it is not apparent from the data displayed to what extent the movement in developed reserves is attributable to purchase and sale. Revisions to reserves generally are not classified between developed and undeveloped reserves. J.S. Herold will have attempted to establish this, and in the event that it was able to do so accurately, its figures should be accurate and more useful than simple finding costs.

Value creation

Figure 5.4 used by BP, shows the extent to which the companies have managed to replace the value lost through production. Its input data is drawn from the SEC reserve value data filed by the companies themselves, which, incidentally, is why Shell is excluded. Shell does not comply with the SEC requirement to file valuation data for its upstream assets, on the grounds that such data is misleading.

The two problems with the BP data are that it is entirely dependent on these calculations, which are misleading indications as to market value, and it also includes the effect of revisions, which, especially over a small number of years, can swamp the value added by exploration activity. Yet one must sympathise with an attempt to represent upstream achievement in terms of value creation, rather than simply additional barrels of oil equivalent.

Finding cost

If the methodological problems were not so enormous, for the large companies at least, value creation would undoubtedly be the appropriate measure. Instead, we would recommend the simplest and easiest, namely to use the addition to proven reserves attributable to discoveries and extensions (excluding revisions), and the amount spent on exploration (not merely that charged to the profit and loss account), averaging both over a reasonably long period of perhaps five years. The resulting figure, expressed in dollars per barrel of oil equivalent, is a reasonable measure of exploration efficiency, ignoring the effect of different companies' optimism on reserve bookings, but a poor proxy for overall upstream value creation.

If finding cost, rather than total value, is to be used as a surrogate for upstream ability, it needs to supplemented by a measure of the success, in volumetric terms, of a company in replacing its production. It is easy enough, after all, for an oil company to improve its finding costs by only drilling its best prospects and slashing its exploration budget. This will reduce finding costs, and may be the correct response to a depressed oil market, but it will almost certainly result in the company failing to replace its reserves, which is the path to long term liquidation. (Given this stark choice, however, liquidation is clearly the best, if not the most often chosen, option for the investor.)

Revisions

In light of the risk concerning finding costs, the issue of what to do with revisions becomes more acute. Oil companies tend to err on the side of caution when making the initial estimates of production on which they justify the decision to develop a field. Once a field is in production, estimates of ultimate recoverable reserves are usually, though not always, revised upwards. Thus, if revisions are ignored, the ability of the company to sustain itself is probably being understated.

On the other hand, reserve revisions to large fields can be enormous in relation to what a company finds in any particular year. Since the object of these comparisons is to compare the success of oil companies in replacing reserves, rather than to derive measures of absolute value, we recommend that the ratio used should again include extensions and discoveries only, excluding revisions, and should be compared with depletion through production, over a run of years. Individual annual statistics, whether for reserve replacement or for finding costs, are almost useless, as the data is too volatile to interpret on a year by year basis.

Downstream performance

Company comparison

Downstream, the simplest measure of performance is that of margin per barrel refined or sold. As with all consolidated figures, however, this can disguise important differences of reason for the discrepancies which emerge. Table 5.6 illustrates comparisons between the profitability of the four international US majors and Shell and BP on the basis of their net profits from refining and marketing, on a LIFO basis, in 1991, per barrel of oil product sold. In BP's case, because the group states its operating results before tax, a notional taxation charge has been applied.

Portfolio influence

What Table 5.6 cannot show, however, is the extent to which the relative performance is attributable to being exposed to relatively more or less profitable geographical areas (which may or may not be a temporary advantage), the extent to which it is attributable to being more exposed to marketing or to upgrading plant (which may equally alter over time in

Table 5.6 *Downstream profitability: net profit per barrel sold in 1991, US$*

BP*	1.06
Chevron	0.40
Exxon	1.47
Mobil	0.98
Shell	1.56
Texaco	1.04

** Taxed at a notional 37%.*

the benefits they confer), and the extent to which it is attributable to better management and more efficient operations.

Clearly, being exposed to geographical areas or operations which confer long term comparative advantages are to a great extent a reflection of successful management, not just good luck, and to that extent are worth a premium to the investor. In addition to the generally shown charts of comparative margins, therefore, we suggest that analysts should pay close attention to the balance of a downstream operation between a company's equity oil production, refining capacity and product sales, to the geographical distribution of its downstream exposure, and to the slate of product which it produces.

A case history

A classic example of the dangers of being wrong footed in this regard is offered by the Shell Oil subsidiary of the Shell group over the period 1988–91. Shell Oil had established a commanding position in the US downstream oil industry over many successful years of trading. It was the largest marketer of gasoline in the USA and was unusual in that it had extensive refining facilities on both sides of the Rocky Mountains (which effectively divide the US oil industry into two almost discrete markets). In the mid-1980s, other US oil companies, notably Amoco, undertook extensive investments in upgrading their refineries east of the Rockies, to permit production of a lighter slate of oil products. Shell Oil decided to upgrade its west coast plant, but to defer similar investments in the east.

The late 1980s saw a considerable tightening of the US gasoline market, partly owing to the introduction of tighter vapour emission legislation, and (ironically) partly owing to the tragic explosion of the catalytic cracker at Shell Oil's New Orleans refinery in 1988, which had the combined effect of tightening the US gasoline market still further, while reducing Shell Oil's own capacity. In 1990 and 1991 severe economic recession followed, which reduced the marketing contribution to overall downstream profits dramatically. In 1992 Shell Oil announced the intention, amid a package of radical restructuring exercises, to add further upgrading units to its refineries, effectively withdrawing from the heavy fuel oil market.

The story is in fact illustrative not only of the problems of resource misallocation downstream, but of the importance to an integrated company of attempting to hedge its overall exposure to risk. Shell Oil in 1991 was highly exposed to natural gas, heavy oil, a refining system with inadequate upgrading capacity and to gasoline marketing. Its surest hedge against the combination of low natural gas prices and wide light/heavy crude oil differentials (which were related by the competition between heavy fuel oil and natural gas in the power generation market) would have been to maintain the highest possible level of upgrading capacity at its refining plant. In fairness, it should be added that the story is also notable as a rare case of significant strategic misjudgement by the Shell group.

Petrochemicals

History

There was a time when many oil companies made large and for the most part extremely unsuccessful diversifications out of their core industry, which was at the time believed to be in inevitable decline (a response to the 1970s 'Club of Rome' fears of the exhaustion of non-renewable resources). These mainly 1970s excursions were largely reversed in the post-1986 determination to sell non-performing assets and reduce debt levels.

Vertical integration

It is now the case that most oil companies have retained little that is significant and saleable outside the oil industry, apart from their chemicals interests. In many cases, these are a direct vertical integration from their refining activities. The chemicals industry is greatly more varied and complicated than the oil industry, but the part of it which is of most importance to most oil companies, the petrochemicals industry, is essentially concentrated on the use of naphtha or LPGs as a feedstock to produce primarily ethylene, and also styrene and propylene, and with the subsequent use of these in the production of polymers (polyethylene, polystyrene and polypropylene). In some cases, the plastics activities of the oil companies extend to the production of finished items for sale to, for example, the automotive industry. In addition to these downstream diversifications, however, a number of oil companies have substantial chemicals activities which are not linked directly to the oil industry. For example, Elf Aquitaine has a substantial US fertiliser business based upon its acquisition of Texas Gulf, a miner of sulphur. It is also a large producer of chlorofluorocarbons, and of ozone-friendly substitutes.

Speciality chemicals

A number of oil companies deliberately extended their chemicals activities into the so-called speciality chemicals market in the 1980s, in the belief that these would offer superior and more stable earnings growth. Speciality chemicals are not well defined, but are generally taken to mean those areas where, because of monopoly of brand management, companies can set their prices independently of commodity price cycles. In general terms, it would appear that the profits which they offered were more stable but were not, over the full business cycle, more profitable. Bulk petrochemicals is an extremely cyclical industry, but the experience of the 1980s suggests that the average return on capital which it offers is no lower than for speciality products.

Comparative advantage

It could be argued (and this is the conclusion which both Petrofina and Shell seem to have reached) that the area in which oil companies might be expected to have a comparative advantage is in fact the bulk petrochemicals area. It can be directly integrated into their refineries, it involves an extension of technology with which they are familiar, and it requires an ability to tolerate violent swings in cash flow, with which the oil companies are all too familiar.

Conclusions

To make a proper operational assessment of the various chemicals activities with which the oil companies are involved would require a complete analysis of much of the chemicals industry, which is beyond the scope of this book. Clearly, the most important measure of success is return on capital employed over time. Useful pointers are offered by the extent to which the chemicals operations are integrated with the refining operations, how far downstream they go, and how volatile the earnings stream has been over time, and whether they offer any cyclical benefits in balancing the overall group portfolio of assets.

In the bulk chemicals industry, as in the refining industry, transport of finished products is expensive, so markets tend to fragment by geographical region. Geographical exposure is thus a vitally important point, especially since demand for plastics is remarkably sensitive to movements in GDP, rising as much as two or three times the rate of overall economic activity at times of rapid expansion.

Finance

Introduction

While comparison of performance and profitability is of great importance in any appraisal of a company, its financial situation should not be ignored. There are three aspects to this:

1 The extent to which the company chooses to fund itself through the medium of equity or of debt.

2 The timing of its cash flows, and of its propensity to be cash absorbent or cash generative.

3 Matters related to dividend distribution.

Financial gearing

On the assumption that there is no fiscal bias (which invalidates the theory somewhat in many economies) an economic theory known after its originators as the Miller-Modigliani theory shows that the value of a company should be entirely independent of the extent to which it is financed by debt or by equity. Increasing the proportion of capital employed represented by debt – the gearing ratio – should have the result of increasing earnings per share, so long as the return on capital exceeds the cost of debt, and also of increasing the return demanded by the shareholders in return for retaining a riskier asset. These two factors should, in theory, cancel each other out to leave the evaluation of the assets unchanged.

Dividend payout ratios

There is also a further Miller-Modigliani theory to the effect that the value of a share should be independent of the proportion of its profits which the company chooses to distribute (again assuming no fiscal distortions). In essence, this is because if the company returns its cost of capital, the investor will be indifferent as to whether the profits are ploughed back into the business, or reinvested by the investor in alternative assets.

Share price rating

Earnings multiple

Share price rating is mentioned earlier but, as stated previously, this is not an unambiguous concept for companies which are not valued with respect to net present value calculations. Most companies' shares are valued with respect to the multiple to earnings at which they trade. As we have seen, net income is not a clear guide to value for oil companies, mainly because of the lumpiness of their cash flows. However, the multiple to LIFO earnings is in general the simplest guide to stock market rating, and perhaps the most useful.

Dividend yield

Dividend yield alone is no guide to value, since (as with other industries) a high yield may simply mean that the company is distributing a large part of its earnings and offers little prospect of growth.

Cash flow multiple

A third measure frequently used in share price determination (especially in the US) is multiple to cash flow from operations. For this purpose, cash flow is taken to be net income plus minorities, depreciation, provisions and other non-cash items. It is a measure of the cash flow available for reinvestment or distribution. Taken alone, however, it is highly misleading if it is not adjusted for debt, on the principle that the value of a bundle of assets is entirely independent of how they are financed.

Table 5.7 shows two comparisons of multiple to cash flow for a group of oil majors. The first simply divides the market capitalisation of each company's equity by its cash flow, to derive an unadjusted multiple. The second adjusts for the effect of gearing by adding back to the cash flow the net interest paid on debt. The net debt is then added to the market capitalisation of the equity, to derive the value which the equity market is, by implication, placing on its unindebted assets. This total value is then divided by the adjusted cash flow to derive a debt adjusted cash flow

Table 5.7 *Debt adjusted cash flow (local currency), m*

	BP	*Shell*	*Royal Dutch*	*Elf*	*Total*	*Repsol*
Price (local currency)	234.5	545.0	147.3	349.5	222.0	2 835.0
Shares issued (million)	5 411.0	3 314.5	536.1	254.1	210.4	300.0
Market cap	12 688.8	18 064.0	78 967.5	88 808.0	46 708.8	850 500.0
Net debt	8 601.0	678.0	2 849.0	31 475.0	14 618.0	140 104.0
Adjusted cap	21 289.8	18 742.0	81 816.5	120 283.0	61 326.8	990 604.0
Cash flow	2 814.0	2 652.0	10 722.0	26 681.0	11 572.0	164 100.0
Taxed interest paid	430.0	34.0	142.0	1 574.0	731.0	8 205.0
Adjusted cash flow	3 244.0	2 686.0	10 864.0	28 255.0	12 303.0	172 305.0
Cash flow multiple	4.5	6.8	7.4	3.3	4.0	5.2
Debt adjusted multiple	6.6	7.0	7.5	4.3	5.0	5.7

multiple. The apparent cheapness on multiple to cash flow of highly geared companies rapidly disappears. The formula for this is as follows:

(Net debt + minorities + market capitalisation of equity)/
(Cash flow from operations + net interest paid)

Conclusion
To conclude, we would suggest the use of two measures for the comparative rating of integrated oil company shares, i.e. multiple to LIFO net income per share, and debt adjusted cash flow multiple. We would strongly suggest that analysts should treat the (commonly quoted) unadjusted multiple to cash flow per share with suspicion.

Strategy

Proved by performance
A final point on comparative analysis is that concerning company management and strategy. To some extent, this is indistinguishable from the operational and financial performances which we have explained how to go about comparing. Good management is one which not only runs competitive operations, but which is more often than not 'in the right place at the right time', and which is sufficiently aware of economic cycles to manage its cash flows so as not to be vulnerable at the wrong time. Luck, however, is also at play, and the analyst has little choice but to decide on almost completely subjective grounds as to how much credit, or otherwise, to give the management and how much to ascribe to fortune.

Stock market circularity
On the stock market it is notable, if unfortunate, that the esteem in which a management is held is often directly correlated with the recent performance of its shares. Companies are usually fairly open about their strategic goals and capital expenditure policies, and these should be assessed by the analyst in the light of his expectations for the industries and regions in which a company operates.

Times of change
There are points in the evolution of a company or an industry when management is confronted by choices which will drive the success of its

business for years to come. Two such which have emerged on either side of the Atlantic in the early 1990s are the issues of environmental expenditure on refineries in the USA and the question of how to react to reduced exploration tax relief for the independent oil companies in the UK. It is at such moments that it is particularly important for the analyst to form a judgement, however subjective, of the strategies which individual companies are pursuing.

ABSOLUTE EVALUATION

Introduction

As will have been noted from the preliminary discussion to the previous section, comparative analysis is probably the most widely used form of appraisal, both in the industry and by investors. It enables companies to assess their performance and to establish targets for their operating companies to beat. And it permits investors to establish views as to appropriate relative rating, and as to the exposure to different categories of risk which the stocks offer.

Industry evaluation

There are times when both industry and investor analysts will want to establish, however tentatively, an assessment of absolute value. In the case of companies, this generally occurs when they are contemplating a bid. Among investors, it is more a matter of style. Some (mainly US) investors will screen the equity market for value, rather than trying to use shares as ways of playing expected future macroeconomic trends.

For the industry analyst who is assessing an acquisition, there is no alternative but to value on an asset by asset basis, upstream and downstream. The putative bidder will model individual oil and gas fields, and will almost certainly try to establish a value for the refining plant from assumptions as to future cash flow and capital expenditure requirements. The marketing assets will probably be valued on the basis of expected margins per litre sold, rather than by individual station. As will have been noted, the evaluation of refineries is far more complex, and more sensitive to external assumptions, than is the case with upstream assets. Furthermore, the strategic considerations of how plant fits into a company's other activities, both in terms of production and distribution, will weigh heavily in its decision as to how much to pay for the assets.

Investment evaluation
For the investor, on the other hand, these factors do not apply. The investor is confronted by the company's shares as an investment opportunity, and his evaluation must be of the company as it currently stands. As pointed out earlier, his returns from the investment will accrue in the form of a stream of income, and his evaluation is therefore a matter of establishing the value of that stream of dividend. This calculation is not affected by his decision to sell the shares later for a capital gain, since the purchaser of the shares will presumably have bought with a view to his future stream of income and growth prospects. An efficient market will always value shares on a 'hold to perpetuity' basis.

Dividend discount models

Construction of a dividend discount model is the standard method of equity evaluation recommended in the economics textbooks, and is used by some investors as the main tool for investment decision making.

Dividend discount models (for all companies, not just oil companies) are highly sensitive to assumptions made about external factors, and the apparent evidence of overvaluation or undervaluation in the equity market which they demonstrate is often largely a function of the assumptions made by the analyst. In addition to uncertainty as to external factors, there is an additional fundamental difference between a dividend discount model for a company and the calculation of a redemption yield for a bond.

Shares have no redemption date, so the analyst has the choice between modelling for a period which is reasonably forecastable and then simply ascribing a residual value to the company at the end of his forecasting period, perhaps five or seven years away, or of modelling on the basis of extrapolation to infinity, in which case the longer term streams of dividend which he projects will be generated by assets which are quite different from those currently owned by the company.

The medium term

Planning cycle forecasts

A pragmatic solution to the problem of time when analysing a company is to attempt to model the consolidated accounts, including a plausible dividend flow, for a company over a period which is long enough to extend beyond the impact of an economic cycle on the downstream and chemicals businesses, while still not requiring extrapolation beyond the planning cycle of the company itself, or the life of most of the assets which it holds at the time of the analysis.

Reasonable extrapolation

An industry buyer would inevitably want to value a target company's assets on an asset by asset basis, and this would probably involve projections over the lives of the assets concerned. For the purposes of investment decisions, lack of resources and data dictate a different approach. Given their maturity, it is reasonable to model companies over a period of five years, and to assume subsequent growth on the basis of normalised returns on capital.

At this point, it would be relevant to highlight Table 5.8, which was constructed by one of the authors for BP immediately after the announcement of a halved quarterly dividend distribution with regard to the second quarter 1992. At the time of the announcement, BP gave some fairly detailed information as to its budgets for cost cutting and for reduced capital expenditures, and as to what it took these to imply in terms of its ability to maintain its levels of production.

Growth rates

Clearly, the implications of Table 5.8 for the earnings and cash flows, which would in reality drive subsequent dividend decisions, would be a function of market forces external to the company. However, Table 5.8 makes what were at the time consensus assumptions as to future oil prices and downstream margins, in an attempt to establish the gradient of medium term growth prospects for earnings and dividends.

Sensitivity

In addition to attempting to establish the shape of the medium term financial outlook for a company, an analysis such as that of Table 5.8 lends itself to the application of a variety of sensitivity analyses to the forecasts; it should be noted however that extreme assumptions regarding the oil price, for example, would have effects on the marginal rate of

Table 5.8 *A five year forecast of BP, £m*

Assumptions	
Price increase (%)	6.0
Inflation (%)	5.0
Dividend (%)	5.0

	1990	*1991*	*1992**	*1993**	*1994**	*1995**	*1996**
Oil production (mboe/d)							
UK	481	479	450	450	450	450	450
US	790	774	650	640	630	580	530
Other	279	313	400	425	450	500	550
Oil equivalent (mbbl/d)	1 550	1 566	1 500	1 515	1 530	1 530	1 530
E and P							
Turnover	6 258	5 761	5 201	5 568	5 961	6 319	6 698
Exploration	−456	−422	−350	−300	−350	−350	−350
Production	−1 268	−1 334	−1 300	−1 336	−1 430	−1 516	−1 607
Production tax	−1 348	−1 001	−731	−832	−879	−940	−1 005
Other	−1 591	−347	−352	−377	−404	−428	−454
Depreciation	−1 101	−1 181	−1 199	−1 284	−1 374	−1 457	−1 544
Abandonment	91	−76	−77	−83	−88	−94	−99
Operating profit	1 585	1 400	1 192	1 357	1 435	1 534	1 639
Production	1 585	1 400	1 192	1 357	1 435	1 534	1 639
Other E and P	501	402	400	400	400	400	400
E and P	2 086	1 802	1 592	1 757	1 835	1 934	2 039

Table 5.8 *Continued*

	1990	*1991*	*1992**	*1993**	*1994**	*1995**	*1996**
E and P (£/bbl)							
Turnover	11.06	10.08	9.50	10.07	10.67	11.31	11.99
Exploration	−0.81	−0.74	−0.64	−0.54	−0.63	−0.63	−0.63
Production	−2.24	−2.33	−2.37	−2.42	−2.56	−2.71	−2.88
Production tax	−2.38	−1.75	−1.33	−1.50	−1.57	−1.68	−1.80
Other	−1.04	−0.61	−0.64	−0.68	−0.72	−0.77	−0.81
Depreciation	−1.95	−2.07	−2.19	−2.32	−2.46	−2.61	−2.76
Abandonment	0.16	−0.13	−0.14	−0.15	−0.16	−0.17	−0.18
Operating profit	2.80	2.45	2.18	2.45	2.57	2.75	2.93
Production tax (%)	46	42	38	38	38	38	38
Refinery throughputs (mbbl/d)	1 779	1 717	1 750	1 750	1 750	1 750	1 750
Refining and marketing profit	853	945	319	511	703	894	1 086
Profit (£/bbl)	1.31	1.51	0.50	0.80	1.10	1.40	1.70
Operating profits							
E and P	2 086	1 802	1 592	1 757	1 835	1 934	2 039
Refining and marketing	853	945	319	511	703	894	1 086
Chemicals	129	−7	50	100	250	300	325
Nutrition	48	37	50	50	0	0	0
Other	−154	−222	−200	−150	−200	−200	−200

Exceptional items	0	0	−1 016	−200	0	0	0
Total	2 962	2 555	796	2 068	2 587	2 928	3 250
Stocks	477	−629	0	0	0	0	0
HCA profit	3 439	1 926	796	2 068	2 587	2 928	3 250
Interest charge	−671	−723	−696	−758	−715	−644	−590
Pretax profit	2 768	1 203	99	1 310	1 872	2 284	2 660
Tax charge	−1042	−820	−497	−590	−805	−982	−1 144
Tax rate (%)	38	68	500	45	43	43	43
Earnings	1 726	383	−398	721	1 067	1 302	1 516
Minority	−50	32	0	0	−10	−20	−30
Net income	1 676	415	−398	721	1 057	1 282	1 486
Dividends	−861	−905	−566	−453	−476	−499	−524
Retained income	815	−490	−964	268	581	783	962
Cash flow							
Operating profit	3 439	1 926	796	2 068	2 587	2 928	3 250
Depreciation	2 011	2 079	2 100	2 000	2 200	2 250	2 300
Exploration charges	460	425	350	300	350	350	350
Associates	−363	−454	−450	−475	−500	−525	−550
Provisions	21	222	0	0	0	0	0
Profit on disposal	−495	−303	0	200	0	0	0
Working capital	−219	345	0	0	0	0	0
Write downs	0	0	544	0	0	0	0
Cash from operations	4 854	4 240	3 340	4 093	4 637	5 003	5 350
Net interest paid	−464	−454	−446	−508	−465	−394	−340
Dividends received	130	141	150	160	170	180	190
Tax paid	−758	−946	−820	−497	−590	−805	−982
Gross cash flow	3 762	2 981	2 223	3 248	3 753	3 984	4 218

Table 5.8 *Continued*

	1990	*1991*	*1992**	*1993**	*1994**	*1995**	*1996**
Investing activities							
Capex	−3 604	−3 793	−3 500	−2 900	−2 900	3 200	−3 500
Acquisitions	−51	−440	0	0	0	0	0
Disposals	1 637	971	800	800	800	600	500
Total	−2 018	−3 262	−2 700	−2 100	−2 100	−2 600	−3 000
Dividends paid	−798	−854	−566	−453	−476	−499	−524
Shares issued	19	16	15	15	15	15	15
Net cash flow	965	−1 119	−1 028	710	1 192	900	708
Balance sheet							
Net debt	6 299	7 437	8 465	7 755	6 563	5 663	4 955
Minority	240	300	300	300	310	330	360
Equity	11 001	10 654	9 705	9 988	10 584	11 382	12 359
Capital employed	17 540	18 391	18 470	18 042	17 457	17 375	17 673

Ratios (%)							
Earn payout	51	218	−142	63	45	39	35
Cash payout	23	30	25	14	13	13	12
Return on capital employed		6	2	8	10	11	12
Return on equity		4	−4	7	10	12	13
Debt: equity	56	68	85	75	60	48	39
Interest rate		7	6	6	6	6	6
Per share statistics							
Shares issued (m)	5 350	5 379	5 393	5 393	5 393	5 393	5 393
Earnings per share (p)	31.3	7.7	−7.4	13.4	19.6	23.8	27.6
Cash flow per share (p)	70	55	41	60	70	74	78
Dividend per share (p)	16.05	16.80	10.50	8.40	8.82	9.26	9.72
Net asset value per share	206	198	180	185	196	211	229

* *forecast*

taxation and (probably) on the profitability of downstream businesses. It is not realistic to change one variable dramatically and assume that the others all remain as before.

Forecasting to perpetuity

Beyond asset lives

An alternative approach to evaluation of shares in integrated oil companies is to apply standardised dividend discount model theory, which requires inputs for a period which extends beyond the lives of existing upstream assets. This requires a different approach, since it will require annual forecasts of upstream profits for a period which will increasingly be influenced by the addition of hitherto undiscovered, not just undeveloped, reserves.

Reserve replacement

A solution to the problem of projecting beyond the life of existing assets is to work from the base of currently developed reserves, and then assume a replacement cost (finding and development cost) per barrel to derive a separate stream of production from newly developed barrels. As is illustrated by Table 5.9, which one of the authors produced for the Royal Dutch/Shell Group in early 1993, this permits the extrapolation of upstream reserves and financial accounts infinitely into the future, the driving assumptions being replacement cost and capital expenditure.

Two simplifying assumptions in the example of Table 5.9 are that the operating costs and tax rates which apply to new oil will be the same as those which applied to the base reserves, and that the exploration charge is capitalised and amortised, rather than written off as unsuccessfully incurred. In effect, a successful efforts company is being modelled as if it were a full cost accounter, but the impact on a mature and stable volume stream will be small, and the alternative would be to split out projections of developed and undeveloped reserves.

Returns on capital

For the other businesses in an integrated oil company, such a model uses fixed assets employed as the basis for projection, projecting return on assets, depreciation, and indexing capital expenditure from an assumed base level. This is simply to avoid complications in the modelling of working capital, and the returns on fixed assets have been regressed towards a figure which equates to a normal return on capital employed. At the corporate level, apart from central overheads, an adjustment must be made for the effect of net cash outflow or inflow on future financial charges.

The residual value

There are a number of approaches to the problem of extrapolation to infinity implied by the construction of dividend discount models. All of them depend on the assumption of convergence to a normal level of profitability over the very long term.

The assumption used in this instance is that a company's growth after the end of the forecast period is determined by the level of return on equity and dividend payout ratio to which the model has converged by the last forecast year. The formula to derive subsequent growth is:

$$\text{Growth} = \text{return on equity} \times \text{proportion of earnings retained}$$

The stream of dividend which accrues to the shareholder after the end of the forecast period can then be valued as an indexed annuity. Its value at the end of the forecast period (which has to be discounted back to a present value) is therefore:

$$\text{Value (end period)} = \text{dividend}/(\text{discount rate} - \text{growth rate})$$

In other words, the residual yield is the discount rate minus the growth rate. Adding together the two net present values, for the forecast dividends and the residual value, derives a net asset value per share. This is, in theory, the price at which the shares should be trading if the assumptions in the model accurately reflect market expectations.

The medium term dividend progression

For the period which the model of Table 5.9 forecasts – ten years of projections – the dividend growth rate has been set so as to be consistent with the residual growth rate which the model derives for the period after the projected period. In other words, the dividend growth rate used to derive a share price value is that which is estimated to be indefinitely sustainable. In the event of the group having a stated dividend policy which conflicts with this, perhaps because it has stated the intention of increasing or reducing its payout ratio, adjustment should be made over the forecast period.

Table 5.9 *Dividend discount model of Shell (forecast)*

Cost inflation, %	5.0
Oil price increase, %	5.0
Dividend growth, %	5.6

Year	*1993*	*1994*	*1995*	*1996*	*1997*	*1998*	*1999*	*2000*	*2001*	*2002*
Developed reserves, mmboe										
Opening base reserves	9 729	8 504	7 433	6 497	5 679	4 964	4 339	3 793	3 315	2 898
Base production	−1 225	−1 071	−936	−818	−715	−625	−546	−478	−417	−365
Closing base reserves	8 504	7 433	6 497	5 679	4 964	4 339	3 793	3 315	2 898	2 533
Base reserves/production	8	8	8	8	8	8	8	8	8	8
Opening additional reserves	0	1 140	2 137	3 008	3 770	4 436	5 017	5 526	5 971	6 359
Additions	1 140	1 140	1 140	1 140	1 140	1 140	1 140	1 140	1 140	1 140
Additional production	0	−144	−269	−379	−475	−558	−632	−696	−752	−801
Closing additional reserves	1 140	2 137	3 008	3 770	4 436	5 017	5 526	5 971	6 359	6 699
Additional reserves/ production		8	8	8	8	8	8	8	8	8
Opening total reserves	9 729	9 644	9 570	9 506	9 449	9 400	9 357	9 319	9 286	9 257
Total production	−1 225	−1 214	−1 205	−1 197	−1 190	−1 184	−1 178	−1 173	−1 169	−1 166
Additions	1 140	1 140	1 140	1 140	1 140	1 140	1 140	1 140	1 140	1 140
Closing total reserves	9 644	9 570	9 506	9 449	9 400	9 357	9 319	9 286	9 257	9 232
Total reserves/production	8	8	8	8	8	8	8	8	8	8
Replacement, %	93	94	95	95	96	96	97	97	98	98
Base $/boe										
Price	16.20	17.01	17.86	18.75	19.69	20.68	21.71	22.80	23.93	25.13
Operating cost	−6.21	−6.52	−6.85	−7.19	−7.55	−7.93	−8.32	−8.74	−9.17	−9.63
Non-cash costs	−4.28	−4.28	−4.28	−4.28	−4.28	−4.28	−4.28	−4.28	−4.28	−4.28
Tax	−3.64	−3.97	−4.31	−4.66	−5.03	−5.42	−5.83	−6.26	−6.71	−7.18
Tax, %	64	64	64	64	64	64	64	64	64	64

Income	2.07	2.24	2.42	2.62	2.83	3.05	3.28	3.52	3.77	4.04
Cash flow	6.35	6.52	6.70	6.90	7.11	7.33	7.56	7.80	8.05	8.32
Base E and P										
Revenue	19 845	18 214	16 716	15 342	14 081	12 923	11 861	10 886	9 991	9 170
Operating cost	−7 607	−6 982	−6 408	−5 881	−5 398	−4 954	−4 547	−4 173	−3 830	−3 515
Depreciation, depletion and amortisation	−5 243	−4 583	−4 006	−3 501	−3 061	−2 675	−2 338	−2 044	−1 787	−1 562
Tax	−4 459	−4 255	−4 034	−3 814	−3 598	−3 388	−3 185	−2 988	−2 800	−2 619
Tax, %	64	64	64	64	64	64	64	64	64	64
Income	2 536	2 394	2 269	2 145	2 024	1 906	1 791	1 681	1 575	1 473
Cash flow	7 779	6 976	6 275	5 647	5 085	4 581	4 130	3 725	3 361	3 035
Additional $/boe										
Price		17.01	17.86	18.75	19.69	20.68	21.71	22.80	23.93	25.13
Operating cost		−6.52	−6.85	−7.19	−7.55	−7.93	−8.32	−8.74	−9.17	−9.63
Non-cash costs		−5.70	−5.85	−6.02	−6.19	−6.38	−6.58	−6.80	−7.03	−7.28
Tax		−3.07	−3.30	−3.55	−3.81	−4.08	−4.35	−4.64	−4.94	−5.26
Tax, %		64	64	64	64	64	64	64	64	64
Income		1.72	1.86	2.00	2.14	2.29	2.45	2.61	2.78	2.96
Cash flow		7.42	7.71	8.01	8.33	8.67	9.03	9.41	9.82	10.24
Additional E and P										
Revenue		2 442	4 806	7 104	9 347	11 547	13 715	15 861	17 994	20 123
Operating cost		−936	−1 842	−2 723	−3 583	−4 426	−5 258	−6 080	−6 898	−7 714
Non-cash costs		−818	−1 575	−2 279	−2 939	−3 564	−4 160	−4 733	−5 288	−5 832
Tax		−440	−889	−1 345	−1 808	−2 276	−2 751	−3 231	−3 717	−4 209
Tax, %		64	64	64	64	64	64	64	64	64
Income		248	500	757	1 017	1 280	1 547	1 817	2 091	2 368
Cash flow		1 066	2 075	3 035	3 956	4 844	5 707	6 550	7 379	8 200
Year end fixed assets	6 500	12 507	18 098	23 344	28 305	33 037	37 588	42 001	46 316	50 568

Table 5.9 *Continued*

Year	*1993*	*1994*	*1995*	*1996*	*1997*	*1998*	*1999*	*2000*	*2001*	*2002*
Total $/boe										
Price	16.20	17.01	17.86	18.75	19.69	20.68	21.71	22.80	23.93	25.13
Operating cost	−6.21	−6.52	−6.85	−7.19	−7.55	−7.93	−8.32	−8.74	−9.17	−9.63
Non-cash costs	−4.28	−4.45	−4.63	−4.83	−5.04	−5.27	−5.52	−5.78	−6.05	−6.34
Tax	−3.64	−3.87	−4.09	−4.31	−4.54	−4.79	−5.04	−5.30	−5.57	−5.86
Tax, %	64	0.05	0.05	0.05	0.05	0.05	0.05	0.05	0.05	0.05
Income	2.07	2.17	2.30	2.42	2.56	2.69	2.83	2.98	3.14	3.30
Cash flow	6.35	6.62	6.93	7.25	7.60	7.96	8.35	8.76	9.19	9.64
Replacement cost	5.70	5.99	6.28	6.60	6.93	7.27	7.64	8.02	8.42	8.84
Total E and P										
Revenue	19 845	20 656	21 522	22 446	23 428	24 471	25 576	26 747	27 984	29 292
Operating cost	−7 607	−7 918	−8 250	−8 604	−8 981	−9 380	−9 804	−10 253	−10 727	−11 229
Non-cash costs	−5 243	−5 401	−5 581	−5 780	−6 000	−6 239	−6 498	−6 777	−7 075	−7 393
Tax	−4 459	−4 695	−4 923	−5 159	−5 406	−5 665	−5 935	−6 219	−6 516	−6 829
Tax, %	64	64	64	64	64	64	64	64	64	64
Income	2 536	2 641	2 769	2 902	3 041	3 186	3 339	3 498	3 666	−3 841
Cash flow	7 779	8 042	8 350	8 682	9 041	9 426	9 837	10 275	10 741	11 235
Capex (including exploration)	−6 500	−6 825	−7 166	−7 525	−7 901	−8 296	−8 711	−9 146	−9 603	−10 084
Net cash flow	1 279	1 217	1 183	1 158	1 140	1 130	1 126	1 129	1 137	1 151
Refining and marketing										
Opening fixed assets	17 500	19 623	21 718	24 024	26 325	28 610	30 892	33 183	35 498	37 849
Return on fixed assets employed	11	12	13	14	14	14	14	14	14	14
Income	1 960	2 355	2 823	3 363	3 685	4 005	4 325	4 646	4 970	5 299
Depreciation	1 877	2 105	2 105	2 329	2 577	2 824	3 069	3 313	3 559	3 807

Cash flow	3 837	4 459	4 928	5 693	6 262	6 829	7 393	7 959	8 529	9 106
Assets/depreciation	9	9	9	9	9	9	9	9	9	9
Capital expenditure	4 000	4 200	4 410	4 631	4 862	5 105	5 360	5 628	5 910	6 205
Closing fixed assets	19 623	21 718	24 024	26 325	28 610	30 892	33 183	35 498	37 849	40 247
Net cash flow	−163	259	518	1 062	1 400	1 724	2 033	2 331	2 619	2 901
Chemicals										
Opening fixed assets	5 625	5 825	6 043	6 300	6 578	6 873	7 187	7 520	7 873	8 246
Return on fixed assets employed	0	5	8	11	15	18	18	18	18	18
Income	−145	291	483	693	987	1 237	1 294	1 354	1 417	1 484
Depreciation	550	570	570	591	616	643	672	703	735	770
Cash flow	405	861	1 053	1 284	1 603	1 880	1 966	2 056	2 152	2 254
Assets/depreciation	10	10	10	10	10	10	10	10	10	10
Capital expenditure	750	788	827	868	912	957	1 005	1 055	1 108	1 163
Closing fixed assets	5 825	6 043	6 300	6 578	6 873	7 187	7 520	7 873	8 246	8 639
Net cash flow	−345	73	226	416	691	923	961	1 001	1 044	1 091
Other										
Opening fixed assets	3 600	3 500	3 426	3 368	3 335	3 325	3 337	3 370	3 421	3 490
Return on fixed assets employed	0	5	8	11	13	15	15	15	15	15
Income	0	175	274	370	433	499	501	505	513	524
Depreciation	400	389	389	381	374	371	369	371	374	380
Cash flow	400	564	663	751	808	869	870	876	888	904
Assets/depreciation	9	9	9	9	9	9	9	9	9	9
Capital expenditure	300	315	331	347	365	383	402	422	443	465
Closing fixed assets	3 500	3 426	3 368	3 335	3 325	3 337	3 370	3 421	3 490	3 575
Net cash flow	100	249	332	404	443	486	468	454	444	438

Table 5.9 *Continued*

Year	*1993*	*1994*	*1995*	*1996*	*1997*	*1998*	*1999*	*2000*	*2001*	*2002*
Group operations										
Net income	4351	5462	6350	7329	8147	8928	9458	10003	10566	11148
Cash flow	12421	13927	14994	16410	17713	19004	20066	21167	22310	23499
Capex (including exploration)	11550	12128	12734	13371	14039	14741	15478	16252	17065	17918
Operating net cash flow	871	1799	2260	3040	3674	4263	4588	4915	5245	5581
Corporate										
Corporate charge	−300	−315	−331	−347	−365	−383	−402	−422	−443	−465
Corporate depreciation	60	63	66	69	73	77	80	84	89	93
Corporate cost	−240	−252	−265	−278	−292	−306	−322	−338	−355	−372
Incremental finance charge	0	−137	−245	−346	−424	−485	−533	−581	−628	−676
Incremental finance cost, %	5	5	5	5	5	5	5	5	5	5
Minority	0	−25	−29	−34	−37	−41	−43	−46	−48	−51
Group net income	4051	4985	5745	6602	7321	8019	8479	8954	9446	9956
Group cash flow	12181	13537	14484	15787	16998	18212	19211	20248	21327	22450
Capex	11550	12128	12734	13371	14039	14741	15478	16252	17065	17918
Dividend	3375	3564	3764	3974	4197	4432	4680	4942	5219	5511
Net cash flow	−2744	−2154	−2013	−1558	−1238	−961	−947	−946	−957	−979
Net debt	3500	5654	7668	9226	10464	11424	12372	13318	14274	15253
Equity (including minority)	50000	51446	53457	56118	59280	62907	66749	70807	75083	79578
Debt/Equity, %	7.0	11.0	14.3	16.4	17.7	18.2	18.5	18.8	19.0	19.2
Payout ration, %	83.3	71.5	65.5	60.2	57.3	55.3	55.2	55.2	55.3	55.4
Return on equity, %	8.1	9.7	10.8	11.8	12.4	12.8	12.8	12.7	12.6	12.6

Table 5.10 *Shell share price value using a variety of exchange and discount rates*

Discount rate, %	£/$ rate	Share value, p 1.60	1.50	1.45	1.40	1.30
11.0		519	586	606	628	676
11.5		501	534	553	573	617
12.0		460	491	508	526	567
12.5		426	454	470	487	524
13.0		396	423	437	453	488

Sensitivities and discount rates

Once the stream of dollar dividends has been derived (Table 5.9 is constructed in dollars, as this permits sensitivity analysis to differing exchange rates), it can be translated into share value in pence using a range of exchange rates and discount rates, as in Table 5.10. Clearly, in addition to looking at the implications of different exchange rates and discount rates, it would equally be possible to change the assumptions on future finding and development costs (upstream) and on normal returns on assets (downstream).

In theory, the appropriate discount rate is the cost of equity to the company, a function of the risk free market return, the equity premium and the beta of the stock. The formula is as follows:

$$\text{Discount rate} = \text{risk free return} + \text{market premium} \times \text{share beta}$$

CONCLUSIONS

Comparison with E and P companies

The points of similarity and dissimilarity between the valuation model which we have discussed for an integrated company and that which we used in calculating asset values for a pure E and P company ought to be noted. In both cases we derive a corporate cash flow model, but in the case of the exploration company, we value the net free cash flows generated by the bundle of assets which the company currently represents; in the other we value the stream of dividends which the company is expected to pay the shareholders after reinvestment in new assets.

Returning cost of capital

In the event that a company returns the cost of capital (its return on capital is identical with returns demanded by the providers of its capital), the present value of the dividend stream will be identical with the present value of the free cash flow generated by its existing assets. This is because the return on reinvested cash flow will be identical with the discount rate used by investors in calculating the asset value.

In theory, under these circumstances, the decision to plough back most of the cash flow, rather than distributing it, will be value neutral to the investor. Instead of the company liquidating itself, it grows its residual value, but the investor will receive only a small proportion of its free cash flows directly in the form of dividend. The reality is that it is highly unlikely that two identical companies, one of which paid out all of its cash flow, and the other of which reinvested all of its depreciation charge and a large part of its earnings, adding up to most of its cash flow (this is of course what most companies actually do) will in fact return the same value to their investors.

Practical considerations

To identify which of the companies in the previous section was the better investment would, however, assume the ability to predict whether or not the company which reinvested its cash flows would be able to generate a better than normal return on incremental investments. In practice, the reason for choosing one method rather than another for the different categories of company is much more basic. Quite simply, it is relatively easy to model cash flows for E and P companies by asset, and very hard to predict the return on incremental exploration expenditure.

For integrated companies, it is virtually impossible to model cash flows by asset, and the range of their activities makes it less unrealistic to think in terms of consolidated returns on capital employed. It has to be admitted, however, that the economic assumptions on which the two approaches are identical are somewhat improbable; this is why, in our discussion of the evaluation of exploration companies, considerable emphasis was placed on the historical and comparative performance of the companies over time, and not merely on their asset values at any given moment.

Theory and practice

The problem of methodology extends to all sectors of the equity market, not just oil companies. The practical difficulty involved with all of the textbook methods of company evaluation are such that equity market practitioners continue to use valuation tools such as multiple to earnings which are thoroughly discredited by economic theory. Their objective is mainly comparative, to compare shares in oil companies with alternatives among other types of investment.

However artificial some of the assumptions behind dividend discount models are, in their attempt to escape from the influence of ephemeral factors on share prices, they should provide a superior guide to value than the traditional rules of thumb.

CHAPTER SIX

*C*onclusions

INDUSTRY BACKGROUND

Industry scale

With the benefit of the chapters on stock evaluation, it should be evident that clear investment conclusions follow from the history and structure of the oil industry. It is in many ways unlike any other area open to equity investment. It is an enormous industry. Taking production of crude oil alone, ignoring the related areas of natural gas, and the shipping, refining and marketing of crude oil and oil products, a world production of about 65 mmbbl/d at a wellhead value of in excess of $15/bbl implies a daily production to a value of over $975 million. The annual equivalent is more than $356 billion.

The pivotal role of oil and gas within the overall world energy mix (between them they accounted for 63% of world fuel consumption in 1991) makes their price a crucial determinant of the cost structure and profitability of manufacturing industries and transport operations.

State ownership

Introduction
Despite these facts, the vast bulk of the reserves of the oil industry are owned by state companies. At the end of 1992, only 8% of world oil reserves were owned by companies in which it was possible to invest. Not only is the bulk of world reserves owned by state oil companies: they own the reserves with the lowest production costs, and the acreage in which there is the best chance of making additional discoveries. Furthermore, they enjoy a surplus of capacity, for it is not their oil which is produced first.

Market manipulation
In a free market it is the cheapest producer who produces first, and the price is set by the marginal cost of the marginal producer. If this were to occur in the oil market then the price would fall, in the short term, to an exceptionally low level, as was demonstrated in 1986. In the longer term, it would rebound violently, since the long run marginal cost from new reserves is much larger than the short run marginal cost from existing reserves.

Since 1986 it has been the decision by the more important state oil producers, i.e. the OPEC countries, to try to balance the market at a price which approximately equates to the long run marginal cost of production which has been responsible for the long period of oil price stability since 1986. It is virtually impossible, at the time of writing, to know what the long run supply curve for the oil industry is, but it is the shape of this curve, and the corresponding set of demand curves for oil products, which establish what is a sustainable price. Recent evidence suggests that costs have fallen considerably in the early 1990s, with corresponding implications for the long run price.

Oil demand

Corresponding to the influence of state agencies on the supply side of the industry, governments in consuming countries are having an increasing effect on demand. While it is in general true that the further downstream one goes in the oil industry, the more assets tend to be owned by privately held companies; the influence of taxation has undoubtedly influenced the behaviour of consumers, and this influence may increase in future. In Europe in particular, the fall in the real oil price since 1986 has not been passed on to consumers, but has been absorbed by national governments.

Analysis of demand for oil is complicated both by taxation and by the fact that it is derived from the demand for individual oil products. The loss, during the early 1980s, of the industrial market in low grade thermal heating (heavy fuel oil) in most of the OECD countries, coupled with the high rate of excise taxes on many oil products, means that the sensitivity of demand for oil to price may now be lower, not higher, than it was in the late 1970s.

Implications for the future

Economic rent

The factors surrounding demand and supply curves suggest that there is an economic rent to oil production, which is related to the cost structure of the source concerned, and that in a market at long term equilibrium it will be a function of the difference between production costs for the producer and the cost of finding and producing new reserves elsewhere. This rent has been transferred from OPEC to consumer governments in the period since 1986, and it is possible that the increasing dependence of consumers on oil from a few OPEC producers will eventually enable them to transfer some of it back again. This may only occur, however, after a period of capacity shortage.

Upstream maturity

The history of the oil companies also has implications for investment decisions, which must be based on views of the future. While the integrated oil companies have mainly arrived at their present shape by vertical integration backwards from a strong downstream oil position (the clearest examples being Shell and the Standard Oil company), the rapid development of independent E and P companies in the USA and in the UK has been associated with the early stage of exploration in what turned out to be a highly prospective basin. E and P companies have been the creatures of – and have ridden profitably – the steep early slope of the creaming curve.

With the potential for significant addition to world reserves of crude oil now much more associated with the application of enhanced oil recovery techniques and the application of high technology to areas from which it has been excluded for political reasons (the Middle East, Latin America and the republics which were once parts of the USSR), the potential to make large returns on capital from the upstream may have swung irrevocably in the direction of large companies. Independents may have the best prospect of making high returns on capital if they are niche players, working reserves and acreage which is of no great significance to the larger companies.

The independents

The constituency of independent oil and gas producers can be divided broadly into three categories: the North Sea exploration companies, the US independents and the Canadian independents (most of which are essentially gas producers). The North Sea companies are now faced by the need to manage a transition from boom to opportunism in a mature province, the role played by the US independents for some years. Only the Canadian gas companies are presented by the possibility of a secular improvement in their market, as the US natural gas price has a long way to rise before it will be capped by the potential imports of LNG from the Middle East or North Africa.

Growth opportunities

One area in which the future does seem to offer the prospect of fragmentation, and the introduction of competitive markets where there have not historically been commercial opportunities, is in the area of gas marketing, especially in Europe. This will almost inevitably result in more volatile prices, but may create opportunities for players other than merely the marketing arms of the major oil companies, who are all diversifying downstream to exploit the opportunity. Moreover, demand for natural gas in both Europe and the Far East is likely to create opportunities for additional supplies of LNG. The main feature of this industry is extremely long time lags and the need for very large capital expenditures.

A second area of growth is the downstream oil industry in the developing countries. Their income elasticity of demand for energy is more than one, and they have little opportunity, in many cases, to substitute away from hydrocarbon fuel sources. This contrasts starkly with the volatile and low returns on capital in the mature downstream industries of the OECD countries.

E AND P COMPANIES

Cash flow modelling

Cash flow modelling of E and P companies is in many ways an ideal evaluative system. By escaping from the use of consolidated accounts and demanding the direct modelling of assets, it permits the evaluation of E and P companies, in effect from the perspective of a corporate bidder. This is all the more significant since these companies are always vulnerable to acquisition by larger, often integrated, competitors.

Buying shares in a company gives an indirect ownership of assets, however, and the cash flows which the assets generate will not flow through to the shareholder unless they are all paid out in dividend (in effect, a liquidation of the company). In practice, it is usually more efficient, not least because of the tax implications, simply to sell the company to a corporate buyer if this were the intention. But in most cases it is not. Companies' managements plough back cash flow in an attempt to add value through new discoveries.

Ongoing entities

Once E and P companies are seen as ongoing entities, then the price which a shareholder should be prepared to pay should be related not merely to the value of a company's existing assets, but also to its perceived ability to add to that value with the cash flows which they plough back. Only companies which return their cost of capital should trade at asset value. If they achieve higher returns on capital, they should trade at a premium; if they achieve lower, then they should trade at a discount. In an efficient market, the latter should eventually be taken over by better management.

While profit and loss accounts may not give much indication of the value of E and P companies (developed assets may produce at a loss, but still have a high value since most of the costs are sunk), they can give a very clear measure of the degree to which they have added to their value by historical trading activity, and this should be an important component in the formation of expectations for the future. This is despite the complications associated with comparing between full cost and successful efforts accounters.

In practice, almost all oil companies give supplementary information in their report and accounts which permits the calculation of fairly accurate figures for their historical success in finding and developing reserves. This offers a more direct measure of achievement.

Strategic problems

It is a function of the size of independent oil companies, especially in the UK, that they are dominated by their stakes in a relatively small number of very large assets. This is less true in the US, where the use of conventional measures, such as multiples to profit or cash flow from operations, is better suited to their evaluation. This feature of the UK companies creates more than merely a problem of evaluation. It represents a strategic problem for the management of the companies themselves, i.e. the extent to which they can really be seen as going concerns.

This issue becomes particularly acute with regard to the issue of payment of dividends. Even when highly profitable, E and P companies are acutely aware of the wasting nature of their asset base. They cannot establish with any certainty the extent to which their cash flows exceed the replacement cost of their fixed assets. Only when a company is very large indeed is it in a position to estimate its replacement costs with any certainty, since the allocation of its exploration budget over a large number of prospects should reduce the volatility of its exploration success.

Cost of capital

With uncertain surpluses, lumpy cash flows and volatile performance, E and P companies represent extremely volatile investments. It is therefore at first sight surprising that their share prices do not in general have high beta coefficients; their cost of equity is not higher than average for the equity market as a whole. The reason lies with the nature of their risk.

The greatest risk to which an investor in E and P companies is exposed is the oil price, yet this is a factor which is highly likely to damage the performance of his other investments, as high oil prices increase energy costs. The reason that the cost of equity to exploration companies is not high is that they represent diversifiable risk from the point of view of equity investors.

Conclusion

Given the uncertainties with regard to performance over time, the best approach to analysis of E and P companies is to refer to a series of performance benchmarks. Asset value is important, but it should not be followed slavishly. Exploration success and the evidence of financial ability to fund its prospects are at least as important.

Lastly, as in the case of all investment analysis of equities, there is the issue of management. The economics of the independent oil companies pose special problems for management, but it is one which as a group they have not addressed very effectively. The problem is how to return value to the investors after an exploration success, and it is complicated by taxation issues. Share prices should, and usually do, reflect opinions on the likely success of managements in achieving this.

INTEGRATED COMPANIES

Scale and diversification

Introduction

While E and P companies are essentially simple companies, with (typically) a small number of discrete assets, each characterised by the need for a high level of development expenditure followed by a long period of positive – albeit diminishing – cash flow, integrated oil companies are among the largest and most complex quoted vehicles on the world's capital markets.

Financial complexity

Analysis of integrated oil companies is not, in general, less sophisticated than that of E and P companies because they are simpler organisations.

It tends to be cruder because integrated oil companies contain so many assets – including the complex and volatile downstream and chemicals businesses – that in practice often the only way to approach modelling them is by extrapolation from published consolidated accounts, rather than by building up cash flows by asset.

The point regarding the complexity of downstream assets is important. Even a very small integrated company is far less amenable to evaluation than an E and P company. The extent to which its downstream assets will require capital investment is uncertain, and its profit margins will typically be highly volatile.

Size and diversification do make the consolidated accounts more useful than they are for small companies, however, especially if the LIFO profit numbers are used. The larger the number of assets which a company owns, the lower the volatility of its performance. In addition, it will also tend to contain assets with a range of maturities, rather than being dominated, as exploration companies often are, by a single asset at a particular state of development or depletion.

Importance of sensitivities

As with E and P companies, analysis of integrated companies should never be static. Whether the methodology used is essentially comparative, or uses dividend discount models to arrive at absolute values, it is vital that the sensitivities of the analysis to changes in assumptions on prices, margins and exchange rates be considered. Very often, what appears to be a valuation anomaly disappears under sensitivity analysis. For example, it is entirely appropriate that the return offered by a share should be relatively high if its profitability is extremely dependent on the profitability of operations in a particular geographical market.

Comparative analysis

A range of benchmarks

Problems of projection, and the way in which many equity portfolios are managed, tend to result in the bulk of professional investment analysis of the integrated companies being essentially comparative in nature. Companies are compared using the benchmarks of operational performance and financial strength, and their share price ratings are compared using a variety of parameters, of which perhaps the most useful is the debt adjusted multiple to cash flow from operations, but also including price to earnings and price to book ratios.

Cash flow most useful

As an example of the parameters that apply when comparing share price ratings, the May 1993 share price ratings of BP, Shell and Royal Dutch with respect to 1992 earnings, dividends and year end asset values are reproduced in Table 6.1; Table 6.2 shows multiples to unadjusted and adjusted cash flows. Exxon's premium is perhaps accounted for by its consistently high profitability. The only surprise regarding the other figures is how tightly bunched they are.

Absolute value

The problem with comparative analysis is that it gives no sense of absolute value, and for the integrated companies this can only be done in terms of dividend discount analysis, or related exercises comparing returns on capital with weighted average costs of capital. Just as an investor should only pay the asset value for an exploration stock which is expected to return its cost of capital, so an integrated company (like any other company) should only trade at its book value if it is expected to earn a return on equity which equates to its cost of equity.

The problem with dividend discount models is that one is required to make highly unrealistic assumptions about profitability from assets which do not yet exist in years which extend beyond the lives of the existing asset base. This is in practice done by assuming a regression to a concept of normalised performance, and the absolute answer

Table 6.1 *Comparison of share price ratings*

	Per share statistics, 1992 local currency			
Group	*Price*	*Price/Earnings*	*Yield, %*	*Price/Net asset value*
BP	307.00	31.0	4.6	1.7
Chevron	84.68	13.0	3.9	2.0
Exxon	65.00	17.1	4.4	2.4
Mobil	69.50	22.4	2.9	1.7
Royal Dutch	164.70	14.2	5.1	1.6
Shell Transport	607.00	18.7	4.8	1.5
Texaco	63.50	18.1	5.0	1.9

Table 6.2 *Adjusted and unadjusted multiples to cash flow from operations*

	Debt adjusted cash flow multiples, 1992 local currency		
Group	*Total capital*	*Price/cash flow*	*Price/debt adjusted cash flow*
BP	26 545	5.4	7.8
Chevron	33 682	7.0	8.0
Exxon	92 639	8.4	9.1
Mobil	36 441	6.9	7.9
Shell Group	54 007	7.6	7.8
Texaco	24 452	6.5	8.1

derived will tend to be highly dependent on the profitability to which the projections regress.

Economic assumptions

To a degree which greatly exceeds the case for E and P companies, analysis of the oil majors requires the taking of macroeconomic views on the outlook for a wide variety of prices and margins, and for economic growth in the geographical markets to which it is exposed. While the crude oil price is set internationally (although that for natural gas, as we have seen, is not) markets for petroleum products vary regionally, and the profitability of marketing petroleum products varies greatly between countries (and in some cases within countries). An obvious example is offered by the comparison between a company which is almost entirely exposed to the mature downstream markets of Western Europe, such as Petrofina, and one which makes some 40% of its petroleum product sales outside Europe and the US, such as the Shell Group. Whatever the fluctuations in margins over the short term, it is hard to believe that the growth prospects for a downstream oil company in Europe can ever match those of the developing world.

Strategic issues

As far as management is concerned, the strategic choices which confront an integrated international major are far greater than those which confront an independent. An integrated company can, through organic change or through deals, shift the balance between the upstream and the downstream, geographically, and in such a way as to lock in margins or not; natural partnerships have been formed by crude short companies and OPEC producers to create joint ventures with downstream assets, for example.

The issue of not merely how operations are balanced but how a group copes with risk extend from the strategic issue of whether to joint venture its downstream oil, to how to hedge its currency exposure or its oil price risk. Growth in derivative markets, both in oil and in financial instruments, has transformed the possibilities for oil company management. It has also made oil company accounts much more opaque, since the nature of a company's real exposure may be far from clear.

EQUITY MARKET BEHAVIOUR

Diversifiable risk

Equity investors invest in companies to obtain a return in the form of capital and income. The return which they demand is a function of the risks which they assume with the investment. Because they look at portfolios as integrated wholes, oil price risk is not a disincentive to invest; indeed, it is by buying oil company equities that investors can hedge their portfolios from the dangers of another oil crisis.

The implication from the point of view of the companies is that their cost of capital is not especially high, and that hedging their crude oil price exposure (as opposed to, for example, refining margins through hedging the spread between crude oil and refined products) will not necessarily make them more desirable investments.

Unlike E and P companies, however, the range of risks to which an international oil major is exposed is vastly greater than simply the crude oil price. Often the most volatile is the spread between crude oil and products, and in this instance their exposure may be quite short term and best locked in by hedging. Other risks, such as economic activity in their downstream markets, are not avoidable, and will affect their cost of equity.

Capital markets

Introduction

Throughout this book, we have talked as if there was one risk free rate of return, one equity market premium and one equity market. The issue of currency has been discussed, but the point was made that much of even an integrated oil company's cash flow is ultimately dollar denominated. In fact, however, oil companies are domiciled in different countries, their shares traded on different markets, and the local currency returns offered by bonds on their local markets are quite different. How this affects their share price ratings and (the other side of the coin) their costs of equity finance is explained in the following section.

International arbitrage

Even if the larger oil companies were not listed and traded on various markets around the world, there would still be a tendency for bond yields to reflect expectations regarding exchange rate movements, so currency adjusted expected risk free returns should not in fact differ between markets so long as there are free flows of capital. This process of adjustment is known as arbitrage. Furthermore, there are sufficient international equity investment flows to make it intuitively implausible that equity premia (based on expected, not actual, returns) vary greatly between markets for particular categories of asset risk. Whatever the historical situation, prior to internationalisation of equity portfolios, it now seems likely that the main implication to the oil company resulting from the market on which its shares are quoted is with regard to liquidity, rather than evaluation.

Liquidity

It is this point which has caused most of the European oil companies to seek listings on the New York Stock Exchange, though they may also have hoped that it would result in a rerating of their shares. It is impossible to prove the point, but it is probably more likely that such convergence as occurred during the late 1980s and early 1990s between the valuations of the continental European oil companies and the US majors was attributable to changes in investor perception of the European companies, rather than a mechanical result of differences between the equity markets; reduction in ownership and control by the state was in several cases accompanied by a markedly more commercial management strategy.

CHALLENGES FOR THE FUTURE

Returning cost of capital

It is of equal importance to the investor and to the corporate groups that the oil companies earn a return on capital which at least equates to their cost of capital during the balance of the 1990s and beyond. Failure to do so would result, ultimately, in poor returns to the investor, and the gradual liquidation of the industry. Shareholder returns can be propped up briefly by increasing dividend payout ratios, but this progression is doomed ultimately to collapse unless the earnings catch up. Equally, it is not necessarily important to shareholder returns whether surpluses are paid out in dividend or ploughed back; the issue is the extent of the surplus.

Other than in a purely cyclical upturn, the parts of industry which tend to offer high returns are those which are expanding rapidly, and in which there is a shortage of capital. Correspondingly, it is extremely difficult to make a reasonable return on capital which is invested in a mature and oversupplied industry.

Implications of capital intensity

The oil industry, because of the scale and complexity of its projects, is one in which there is a long time lag between an economic change and the industry fully reflecting its impact. Competitive advantages and disadvantages tend to be long lasting and hard to overcome. This makes it particularly important for the investor to base his stock selections on an assessment of the economic outlook for the industry, especially those areas in which there seems to be a reasonable prospect of obtaining relatively high returns on capital over a number of years. Where companies are inadequately exposed across these areas, it will be a lengthy and expensive process for them to redress this.

Possible developments

On the current evidence, it may be that it will be hard to generate consistently high returns on capital in mature upstream oil provinces (especially without ownership of the existing pipeline and platform infrastructure), and from the downstream oil and chemicals businesses in the western OECD countries. This does not preclude cyclical fluctuations in the margin businesses.

Areas which seem to offer better than average returns are, perhaps, all constituent parts of the natural gas industry, and the margin businesses in parts of the world in which there will be a strong underlying growth in demand. The most unpredictable element is the possibility that political change will permit high equity returns to re-entrants into the upstream oil industry in low cost countries.

Need for size

All of the trends highlighted in this final chapter, if they have been identified correctly, seem to suggest that the future will be brightest for those oil companies which are already the largest, the most international and which have the strongest balance sheets. In the oil industry it has rarely been the case that 'small is beautiful', and this, at least, seems unlikely to change.

*B*ibliography

CHAPTER 1

British Petroleum, *Statistical Review of World Energy*, 1992.

Cambridge Energy Research Associates, *World Oil Trends*, 1991.

Glennie, K.W., *Introduction to the Petroleum Geology of the North Sea*, Blackwell Scientific Publications, 1984.

Halbouty, M.T., 'Basins and new frontiers: an overview'. In: 'Future Petroleum Provinces of the World', *AAPG Memoir*, no. 40, 1986.

Holmes, Sir P., 'Oil companies' strategies for the 1990s'. Address to the London Oil Analysts' Group International Petroleum Conference, London, June 1988.

International Petroleum Encyclopaedia, Energy Group, Pennwell Publishing Company, Tulsa, Oklahoma, 1992.

'Worldwide production report', *Oil and Gas Journal*, vol. 90, no. 52, 1992.

Organisation of the Petroleum Exporting Countries, *Facts and Figures: A Graphical Analysis of World Energy Up to 1991*, Vienna, 1992.

Peebles, M., 'The role of LNG in the world energy scene', a presentation on behalf of LASMO plc, November 1992.

PIW, *PIW Ranks the World's Top 50 Oil Companies*, December 1992.

Schurr, S.H. and Netschert, B.C., *Energy in the American Economy 1850–1875*, John Hopkins Press, Baltimore, 1960.

Yergin, D., *The Prize*, Simon and Schuster, New York, 1991.

CHAPTER 2

Banks, F., 'Oil now and later', *OPEC Bulletin,* February 1992.
British Gas, *Financial and Operating Statistics,* 1991.
British Petroleum, *Statistical Review of World Energy,* 1992.
Hotelling, H., 'The economics of exhaustible resources', *Journal of Political Economy,* no. 39, 1931.
Mlotock, P., '"$19/bbl oil is not an aberration – It's a return to the norm"', *Middle East Economic Survey,* 23 March 1992.
Odell, P., *Oil and World Power: Background of the Oil Crisis,* Viking Penguin, New York, 1986.
'Market review', *OPEC Bulletin,* September 1992.
Seymour, I., 'OPEC in the 1990s', *Energy Policy,* October 1992.
Yergin, D., *The Prize,* Simon and Schuster, New York, 1991.

CHAPTER 3

Accounting Standards Board, *Financial Reporting Exposure* (Draft 4), 1993.
Antill, N., 'An analyst's view of oil accounts', *Oil and Gas Finance and Accounting,* Spring 1991.
Arthur Andersen, *Financial Reporting by the European Oil Industry,* London, 1991.
Brealey, R. and Myers, S., *Principles of Corporate Finance.* McGraw-Hill, New York, 1982.
Holmes, G. and Sugden, A., *Interpreting Company Reports and Accounts,* Woodhead-Faulkner, Cambridge, 1990.
Stewart, G.B., *Stern Stewart Corporate Finance Handbook,* Stern Stewart Management Services, New York, 1986.
Oil Industry Statements of Recommended Practice:
Disclosure draft oil and gas exploration and production activities, April 1986.
Accounting for oil and gas exploration activities, December 1987.
Accounting for abandonment costs, June 1988.
Accounting for various financing, reserves and other transactions of oil and gas exploration and production companies, June 1991.

CHAPTER 4

Haun, J.D. (Editor), 'Methods of estimating the volume of undiscovered oil and gas resources', *AAPG Studies in Geology,* no. 1, 1975.
Megill, R.E., *An Introduction to Risk Analysis,* Pennwell Publishing Company,

Tulsa, Oklahoma, 1979.

Megill, R.E., *An Introduction to Exploration Economics*, Pennwell Publishing Company, Tulsa, Oklahoma, 1979.

Megill, R.E. (Editor), 'Economics and the explorer', *AAPG Studies in Geology*, no. 19, 1985.

Nederlof, M.H., 'Estimation of undiscovered hydrocarbons – methods and pitfalls', Sixth Exploration Seminar of the Egyptian General Petroleum Corporation, Cairo, 1982.

Newendorp, P.D., 'Petroleum exploration economics and risk analysis', Joint Association for Petroleum Exploration Courses (JAPEC), course notes, no. 93, November 1989.

Newendorp, P.D., *Decision Analysis for Petroleum Exploration*, The Petroleum Publishing Co, Tulsa, Oklahoma, 1975.

Parker, J.R., 'The impact of appraisal methods on exploration decision making', Address to the Norwegian Petroleum Society, Stavanger, 11–12 April 1984.

Schulyer, J.R., 'Modelling an exploration program: Insensitivity to prospect ranking', *Oil and Gas Journal*, vol. 87, no. 52, 1989.

Schulyer, J.R., 'Decision rules using the EMV concept and attitude towards risk', *Oil and Gas Journal*, vol. 88, no. 2, 1990.

Sluijk, D. and Nederlof, M.H., 'World-wide geological experience as a systematic basis for prospect appraisal'. In: 'Geochemical Basin Analysis', *AAPG Memoir*, no. 35, 1984.

Van Dam, J., 'Economic analysis of oil and gas projects', Offshore Technology Course (1980–1981), Delft University of Technology, 1980.

CHAPTER 5

Arthur Andersen, *Financial Reporting by the European Oil Industry*, London, 1991.

Brealey, R. and Myers, S., *Principles of Corporate Finance*, McGraw-Hill, New York, 1982.

British Petroleum, *Our Industry Petroleum*, 1977.

Duggan, J., 'Analysing the strategies of the international oil "Majors"', *OPEC Bulletin*, September 1992.

Holmes, G. and Sugden, A., *Interpreting Company Reports and Accounts*, Woodhead–Faulkner, Cambridge, 1990.

Institute of Petroleum, *Retail Marketing Survey 1993*.

Petroleum Intelligence Weekly, special supplement, 17 August 1992.

Rappaport, A., 'The affordable dividend approach to equity valuation', *Financial Analysts Journal*, July/August 1986.

Rose, I., 'Petroleum retailing – hypermarkets', *Petroleum Review*, June 1992.

Sampson, A., *The Seven Sisters: The Great Oil Companies and the World They Created*, Coronet, London, 1988.

Stewart, G.B., *Stern Stewart Corporate Finance Handbook*, Stern Stewart Management Services, New York, 1986.

Glossary

Abandonment: The point in time when all economic reserves have been extracted from an oil and gas field and production is stopped.

Acreage: An area of land or sea where a company is entitled to explore for oil or gas reserves.

Additional profits tax: A tax designed to generate extra revenue during periods of high oil prices.

Alkylation: A process which involves the polymerisation of propane and butane (LPGs) into iso-octane.

Appraisal: Drilling or seismic activities which aim to delineate the range of reserves made after an exploration well has first encountered hydrocarbons.

Back-in: The participation of a national oil company in an oil or gas development once an exploration success has been made ('state back-in').

Basin: A geological province on land or offshore where hydrocarbons are generated and trapped.

Bitumen: The name applies to the heavier end of the hydrocarbon molecular compounds that exist in crude oil, e.g. asphalt.

Blending: The process of mixing different crude oil types in a refinery.

Build-up period: The period of rising production once a new oil or gas field has been brought on stream.

Capex: Capital expenditure (required to build new facilities).

Carbon taxes: Taxes specifically aimed at limiting or relating to the burning of fossil fuels.

Carried interest: An interest in an oil or gas field whereby all capital expenditure is paid by an additional partner, who recovers the cost from the revenues from production.

Catalytic cracker: A unit in a refinery which uses catalysts to convert a greater percentage of the original feed stock into a higher value product.

CCGT: Combined cycle gas turbine used to generate electricity from gas.

Core asset value: The net present value of commercial and potentially commercial assets.

Cost oil: The portion of oil revenues which is used by companies to recover capital expenditure in production sharing contracts.

Crack: The process by which complex hydrocarbon molecules are broken down into simpler molecules.

Creaming curve: A graphical representation of the cumulative volume of oil found over time in any particular area.

Crude distillation unit: The basic unit which distils crude oil and separates out the main fractions of oil, e.g. distillates from residual oil.

DCF: Discounted cash flow.

Deferred tax: An element of tax in a profit and loss account which provides for future tax liabilities.

Depletion charge: The charge in a profit and loss account which relates to the depreciation of fixed assets.

Development costs: Capital expenditure required to bring oil and gas reserves on production.

Discount factor: The factor by which a value to be received at a future date has to be multiplied to arrive at the present value.

Domestic supply obligation: A volume of oil or gas which a company is obliged to deliver into the local market (usually in production sharing contracts).

Downstream: All activities relating to the refining and marketing of crude oil and gas.

EMV: Expected monetary value.

EOR: Enhanced oil recovery.

Equity crude: The portion of crude oil or gas to which a company is entitled under the terms of a production sharing contract.

Expectation curve: A graph used to express the range of reserves and associated risks that may occur in any hydrocarbon accumulation.

Exploration prospect: A geological structure which may contain hydrocarbons but which has not been tested by any exploration well.

Field depletion contracts: A contract which guarantees the sale of all the oil or gas reserves from any one oil or gas field.

Floating storage: Oil tankers which are used to store crude oil.

Fossil fuels: All fuels formed or generated from organic matter, e.g. oil, gas, coal.

Gasolines: A low boiling petroleum distillate, e.g. petrol.

Hydrocracker: A high pressure and high temperature unit in a refinery which is used to upgrade products, e.g. gas oil to gasoline.

Hydrofiner: A unit in a refinery which uses hydrogen to reduce the sulphur content of the feed stock.

Hydrogen conversion unit: An expensive unit in a refinery which can utilise almost any feed stock and crack it into a relatively light range of distillates.
IEA: International Energy Agency.
Integrated companies: Companies which are involved in both downstream and upstream activities.
Isomerisation: An efficient process which produces high octane gasoline from naphtha.
Joint venture contracts: Contracts for the joint participation in the exploitation and production of oil and gas assets.
Licensing rounds: The allocation of exploration acreage to oil and gas companies in return for a signature bonus, or a minimum level of expenditure, or an agreed price.
LNG: Liquefied natural gas.
Minimum economic volumes: The minimum volume of oil or gas reserves that are required to be discovered in order for a development to be declared commercial.
MOD: Money of the day.
NGL: Natural gas liquids.
Net to gross acreage: The net area of acreage that an oil or gas company owns after taking into account its equity interest.
NYMEX: New York Mercantile Exchange.
NPV: Net present value.
Oil allowance: The volume of oil which can be produced which is not liable to any taxation.
Oil majors: Large integrated companies, e.g. Shell, Exxon etc.
Operating costs: Costs associated with running facilities and producing oil and gas.
OPEC: Organisation of the Petroleum Exporting Countries.
Opex: Operating costs.
Pay-out time: The time required for an oil and gas field to return the value of the initial investment.
Peak shaving capacity: The maximum production capacity of the facilities installed on a gas field, which is typically only utilised during periods of high customer demand, e.g. winter.
Plateau period: The period in the life of an oil or gas field when production is at a peak and level for several years.
Play: A geological concept which may result in the discovery of hydrocarbons if specific exploration prospects can be identified.
Possible reserves: The volume of hydrocarbon reserves that geological, geophysical and engineering data indicate to be in place, for which there is a greater than 90% chance that it will be less.
Present value: The present value of any amount of money to be received at a specified future date equals the amount that would have had to be put on compound interest at the reference date to yield the required amount at the future date.
Primary distillation: The initial process in a refinery where crude oil is separated into separated component parts.

Probable reserves: The volume of hydrocarbon reserves that geological, geophysical and engineering data indicate to be in place, for which there is a 50% chance that it will be more or less than that amount.

Production forecast: A forecast of the future production levels of production that can be expected from any particular oil or gas field. These are often subject to frequent revisions.

Profit oil: The volume of oil in a production sharing contract which remains after the contractor has recovered its costs; the profit oil is shared between the national oil company and the contractor in agreed proportions.

Proven reserves: The volume of hydrocarbon reserves that geological, geophysical and engineering data indicate to be in place for which there is a 90% chance that it will be more than that amount.

PRT: Petroleum revenue tax.

PSC: Production sharing contract.

ROR: Rate of return.

Recoverable reserves: The volume of reserves that can actually be produced from an oil or gas discovery in contrast to the volume of oil that may actually be in place in the ground.

Reserves revisions: A revision (up or down) made to the original estimate of recoverable reserves in an oil or gas field.

ROC: Return on capital.

ROE: Return on equity.

Royalty: A form of tax whereby a host government takes a share of the production in kind or – more commonly – in cash.

Signature bonus: The initial payment made to a host government in order to secure the right to explore or develop the mineral resources in a particular area.

Spot prices: The daily price of crude oil.

UKCS: United Kingdom Continental Shelf.

Upstream: All activities relating to the exploration and production of hydrocarbons.

Vacuum distillation unit: A unit in a refinery which distils the hydrocarbon residue which did not boil at atmospheric pressure in the crude distillation unit.

Visbreaker: A simple conversion process which relies solely on heat to crack the hydrocarbon molecules.

SIMPLIFIED CONVERSION UNITS

Oil

1 bbl	=	0.159 Sm3
1 bbl	=	5800 scf of gas
1 mmbbl	=	5.8 billion scf of gas
1 Sm3	=	6.29 bbl
1 tonne	=	7.4 bbl

Gas

1 scf	=	0.0283 Sm3
1 Sm3	=	35.3 scf

*I*ndex